Bireme warship, relief from the Temple of Fortuna Primigenia at Palestrina (Praeneste), Lazio Region, Italy (*De Agostini/Getty Images*)

THE ROMAN NAVY

SHIPS, MEN & WARFARE 350 BC–AD 475

MICHAEL PITASSI

Seaforth
PUBLISHING

For Ann
my wife and muse

Copyright © Michael Pitassi, 2012

First published in Great Britain in 2012 by Seaforth Publishing
An imprint of Pen & Sword Books Ltd
47 Church Street, Barnsley
S Yorkshire S70 2AS

www.seaforthpublishing.com
Email info@seaforthpublishing.com

British Library Cataloguing in Publication Data
A CIP data record for this book is available from the British Library

ISBN 978-1-84832-090-1

Printed and bound in China through Printworks International

CONTENTS

PREFACE

THE ROMAN NAVY or, more accurately, the naval forces of the Roman state, is more often than not consigned to little more than a couple of paragraphs in many accounts of Roman military endeavour. In fact, it was for over 800 years an integral part of the armed forces of that Roman state and became the world's first 'super-power' navy. It was the instrument by which Rome achieved domination of the western Mediterranean, which enabled her expansion into the lands surrounding it and the foundation of her empire; it was naval forces that enabled the Romans to intervene and eventually to dominate the Eastern Mediterranean and the lands of the near east. It was a naval campaign and a sea battle that established and secured power for the first of the emperors of Rome; it was the navy's domination that enabled trade and the economy of the empire to grow and to flourish, free from the scourge of piracy and to an extent not equalled until the twentieth century. Finally, it was the loss of that domination that was a vital factor in the disintegration and ending of the Western Roman Empire. Cicero's dictum that 'The master of the sea must inevitably be master of the empire' (M. Tullius Cicero, 106–43 BC, *Epistolae ad Atticus* 10.8.4, *existimat enim qui mare teneat eum necesse [esse] rerum potiri*) has proved to be true in every Europe-wide war since. At its height, during the Punic Wars, the navy accounted for fully a third of the total Roman military effort and during imperial times it averaged rather more than a tenth of it.

Surviving evidence relating to the service is sparse and fragmentary but there is overall a body of information about the Roman navy, seafaring and shipping in the ancient world and aspects of Roman life which can be searched to extract a picture of the Roman navy as a whole. In seeking to arrive at a detailed picture of the service, parallel evidence will also have to be relied upon, as will therefore, the argument that it would be unreasonable as well as illogical to suppose that entirely separate arrangements, systems or methods had been employed just for the navy. For example, while on the one hand it seems likely that the army's standard issue sword, the *gladius hispaniensis*, was also used by the marines of the navy, on the other it seems that the navy adopted a short pike for use in boarding which finds no parallel with the army. Conclusions should not, however, be reached without consideration of the appropriateness of an item. So, for example again, it is well known that the Roman legionary wore sturdy, ventilated boots (*caligae*) studded with hobnails, but this footwear would have been impractical for a marine, as the hobnails would lack grip on a wooden deck and would quickly rip the surfaces to shreds. One can thus posit that a marine's boot, although similar, would have substituted a leather or rope sole; the army's boot is proven, the marine's surmised, but well within the technology of the time but there are no surviving examples.

There is some evidence, however, such as inscriptions, wall and pottery paintings, statuary and literary references, that directly relate to the navy. As an example, a votive altar set up by Valerius Valens, in which he describes himself as *Praefect[us] Classis Mis[enensis]* (CIL X.3336/ ILS 3765), prefect of the fleet at Misenum, shows that the post existed in the early third century AD, the date of the altar. There are other traces, too, on the monumental scale; at Carthage, for example, there are the remains of the basins that formed the once great military and mercantile harbours, including the base of an excavated ship shed on the area now known as Admiralty Island. Syracuse has some remains of city walls from ancient times as well as the great Euryalus fortress, the world's first artillery fort. In Rome the Capitoline Museum houses the plaque erected to celebrate Duilius'

The view from Baia (Baiae) to Pozzuoli (Puteoli). Tiberius' philosopher, Thrasyllus taunted Gaius (Caligula) that he could no more become emperor than ride a chariot between them. When he did become emperor, Gaius ordered the nearby Misene fleet to gather every ship that they could find to bridge the distance of some 2 miles (3 km); they built a pontoon bridge from some 2,000 ships, 10 feet (3 m) wide to join them, with 'islands' every thousand paces upon which 'villages' were constructed. Gaius duly rode his chariot across, accompanied by cavalry and infantry, returning across on the following day. (*Author's photograph*)

naval victory at Mylae in 260 BC. At Ostia, by the Marine Gate, the funerary monument of C. Cartilius Poplicola (first century BC) commemorating his military achievements in life is crowned by a representation of a warship of trireme configuration; almost opposite this is a model in marble of a ship's ram. There are smaller artefacts, such as bricks bearing stamps of the fleets that had them made, metal diplomas awarded to men upon their retirement from the service and inscriptions and reliefs on monuments and stelae in many museums, as well as wall paintings, such as those from Pompeii.

Although having had great numbers of ships and men, the Republican navy did not have the great naval bases of later Imperial times and thus little trace remains of it. The Imperial navy did, however, and archaeological excavations have been carried out at many of them, although there is sometimes little to distinguish the remains as distinctly 'naval' as opposed to military. Thus a military hospital unearthed at Xanten (Vetera,

Belgium) could serve both the Rhine fleet and local garrisons. One location known to be of naval significance is, of course, Misenum, much of which is now a pleasant, modern holiday town; but there are visible remains of masonry and moles and the base of the temple of the Augustales, the priests of the cult of the emperors, can be seen, the facade of which is in the nearby Bacoli museum. The great cistern can also be visited. At Ravenna, much less remains: a piece of wall and the remains of a mole, now inland. This site is a good example of how in many places changes of shoreline and silting over the centuries mean that the *raison d'être* and context have been lost. This is not always so, as, for instance, Portchester Castle (Porta Adurnis) in Hampshire, the best-preserved of the 'shore forts', still stands on the shore at the top of Portsmouth harbour, connected to the nearby Roman road network. The piers, jetties, slipways and quays that once served for the operation of its warship complement have long

The trireme *Olympias*, a full-size reconstruction of a fifth century BC Athenian trireme, built in 1987. This type was the ultimate ramming vessel and remained in widespread service even after it had been replaced in the battle-line by larger types. The ship is now at the Averof Museum, Nea Faliro, near Athens. (*Author's photograph*)

since disappeared but can be readily envisaged, the context remaining valid.

For the Romans numismatic evidence is of a useful source of information, as they used warships or the prows of them as designs for coins. Although necessarily stylistic to fit within their tiny compass, these can reveal useful indications as to the appearance of ships and have the benefit of being dateable. Underwater archaeology has discovered and documented many wrecks of merchant ships but, for the reasons given later, no seagoing warships, despite many hundreds being lost. The particular conditions of mud flats and river banks has revealed some examples of warships, dumped, locked in position and preserved by drifting silt. Thus the remains of two Carthaginian ships have been discovered in Sicily and river warships recovered from both the Rhine and Danube and even more recently, possibly from the estuary of the River Arno in Italy.

The body of material relating to Roman naval forces is widely scattered and often lacks continuity; nevertheless there is enough to form a reasonably detailed appraisal of those forces over the many centuries of their existence, which this book set out to do. The current names of places have been used throughout, with their Roman names added at the first instance of each, save for Misenum, the great naval base that encompassed so much more than the present small town; for this and any other idiosyncracies, as indeed for any errors, *mea culpa*. Wherever possible, contemporary evidence has been sought and efforts made to avoid the assumption of conclusions not so supported, any such again being the fault of the author alone.

The Romans are masters of the sea whatever lands they come to, they at once subdue. LIVY, XXXII.21

PART I INTRODUCTION

1 A BRIEF HISTORY OF THE ROMAN NAVY

The early period

Tradition says that Rome was founded on 21 April 753 BC as a monarchy and indeed archaeological finds have confirmed that there were settlements on the hills of Rome from the eighth century BC. The city was located at a point on the River Tiber (Tiberis) navigable by the small and medium-sized ships of the time; the river ran higher then than it does now. The first expansion of Roman power was to control the mouth of the river, 14 miles (24.5 km) away, to secure access to the sea. The last three kings of Rome were Etruscans and marked a century of Etruscan domination. The Etruscans had long been a seafaring people[1] and it is safe to assume that some of the Romans also owned, operated or served aboard ships and that it is from this early time that the first Roman ships

are likely to date. Warships would have been the preserve of the Etruscans although Roman personnel must have served on them as well. In 510 BC the Romans expelled the last of their kings and formed a republic. One of the fledgling state's first acts was to sign the first of three treaties of friendship with Carthage, the pre-eminent maritime power of the central and western Mediterranean.[2]

The earliest record of the existence of a Roman warship is from 394 BC when a warship was sent with three senators and a votive offering to the oracle at Delphi in Greece. The ship was captured by the men of the Lipari (Liparae) Islands, who mistook it for a pirate. Learning of the mistake, the islanders gave the Romans hospitality and escorted them on their way.[3] The Roman ship must have been small to have

An impression of an eighth-century BC warship from a large Greek krater and contemporary with the founding of Rome and of a type operated by the Greeks, Etruscans and also possibly by the Western Phoenicians (later Carthaginians), if not by the Romans themselves. The earliest warships to become known to Rome were of this type, namely, monoreme conters with a pointed ram and a fighting deck forward and a stern deck for the helmsman and captain. This example may be decked along its entire length, judging by the figures shown standing on top. Below that is a lattice of one horizontal wale and seventy-four vertical stanchions; whether this indicates mere decoration of the hull side, or indicates an open, lattice-work along the hull sides, is not clear, but they are too numerous to indicate only thole pins. A prominent *oculus* is on the bow. Attic Greek krater, Metropolitan Museum of Art, New York. (*Author's photograph*)

Capo Colonna , the Lacinian promontory south of Crotone in the western Gulf of Taranto. In the treaty of 338 BC between Rome and Taranto this cape was made the boundary between them and Roman warships were to remain to the west of it (to the right of the picture). It was the breaking of that treaty in 282 BC that led to the Pyrrhic War. The site was famous for and dominated by a temple to Hera Lacinia. The scaffolding tower to the left of the lighthouse encloses the one column of the temple that still stands, and gives it cape its modern name, The remainder of the temple complex lay were between it and the modern buildings to its left. (*Author's photograph*)

been so easily taken and one of only a few such ships then in service, no more than a few ships being necessary for the short length of Roman coast. To demonstrate the nominal size of the fleet at that time, no naval response was made when, in 349 BC, a force of Greek buccaneers cruised up the Italian coast; instead the Romans sent troops to shadow them on shore and to prevent any landing. Unwilling to face Roman troops, the pirates left.[4]

During the Latin War (340–338 BC) the Romans deployed a fleet which gained them their first naval victory against the Latins and Volscians in a battle off Anzio (Antium). The rams (*rostra*) of some of the captured enemy ships were taken to Rome and mounted on the speaker's platform in the Forum, which was subsequently itself call the Rostra.[5] The Greek maritime city of Taranto (Tarentum) made a treaty with Rome in 338 BC, seeking to limit Roman naval activity in the Gulf of Taranto, even though Roman territory had yet to reach that far. The Roman navy was still small, perhaps of twenty ships and those of the smaller types, although the acquisition of a trireme or two, even if only for prestige is feasible.[6] With the absorption of the Greek Italiote colonies (the Greek settlements in southern Italy), the Romans acquired their warships, crews and

shipyards. Such city-states as Naples (Neapolis, acquired in 326 BC) were made exempt from liability to provide troops, but had to provide warships and crews to patrol and guard their harbours and lands.

In 312 BC the Romans seized the offshore Pontine (Pontiae) Islands to protect the coastal route between Ostia and the Bay of Naples.[7] In 311 BC naval forces had become such a regular part of the military establishment that in that year, a 'navy board' of two senior officers known as *duoviri navales* was set up with responsibility to build or acquire sufficient warships and to oversee their equipping, victualling and readiness for service. They were also charged with the provision of and training of adequate crews, in short, to set up and run a navy. At this point the navy had in fact, become a separate body, rather than little more than an ad hoc floating part of the army.[8]

During the Second Samnite War, in 310 BC, Roman naval ships attacked Pompeii (then by the sea) and then moved further inland than was prudent and were counter-attacked, forcing them to flee back to their ships.[9] Modest growth of the service, alongside the growth of the Roman state, continued with varying degrees of success but at a modest level of activity. To a core of Roman-built and manned ships could be added ships, crews and equipment levied from the maritime allies such as Naples, Anzio and Paestum to augment their strength. A further treaty was concluded with Carthage in 306 BC[10] that sought to avoid potential conflict by recognising spheres of influence. Thus Carthage undertook to stay out of Italian affairs and Rome to keep out of Sicily. They further agreed to divide the former Etruscan island of Corsica between them and although the Carthaginians occupied their part, the Romans did not.

Roman hegemony gradually extended across the Italian peninsula and brought them into conflict with Taranto. In 328 BC a Roman squadron of ten ships was in the Gulf of Taranto. The Tarentines manned between fifteen and twenty of their ships and attacked the Romans, sinking four, capturing one and scattering the rest, inflicting upon the Romans their first naval defeat.[11] The five ships lost represented a large proportion of the Roman fleet of the time, which probably numbered about forty ships in all. This estimate arises from an incident in 295 BC, when an attempted plot by 4,000 Sabellian levies for the fleet was foiled. These men were rated as *socii navales* (naval allies) and at an average crew of 200 men for a trireme, would have been enough to man twenty such ships. Accepting that these (unreliable) men were allowed to be less than half of the service's manpower, forty triremes could be crewed. However, many of the ships would be smaller than triremes and a few bigger, so the number is posited as reasonable.[12]

With the end of the Pyrrhic War in 275 BC and the surrender of Taranto, the Romans had extended their territory over the whole Italian peninsula, with a consequent increase in coastline and in the need for naval forces. The further growth of Roman power and influence, as well as of her trade made it apparent by the early third century BC that a trial of strength over who was to be the dominant power in the central and western Mediterranean, between Rome and Carthage would become inevitable.

The Punic Wars

Further Roman expansion was limited by the existence of Carthage and her empire. This included, as well as part of Corsica, a large part of Sardinia, the western two-thirds of Sicily, much of the North African coastal area of Tunisia and Algeria and a part of southern Spain. It was sea-based trading empire, jealously guarding the sea routes and restricting trade and navigation by all others. To enforce this policy, Carthage had a huge navy with its own purpose-built military harbour at Carthage, capable of supporting up to 200 warships. The navy was a standing force, well organised and practised.[13]

War finally broke out in 264 BC over possession of Messina (Messana) on Sicily, a war that would essentially be for possession of that island. The Punic fleet there left Messina, allowing a Roman force to cross the strait from the mainland, eject the Punic garrison and occupy the city. Denied its harbour, the Punic fleet had to fall back on their next nearest harbour, Milazzo (Mylae) on the north coast of Sicily and try to blockade the straits from there, to prevent the Romans from crossing in strength. The strait is long (24 miles, 39 km) and narrow (2.5 miles, 4 km) deep and subject to sudden squalls, with stony beaches, strong currents and difficult to patrol.[14]

The Punic fleet made raids upon the Italian coast but nevertheless, by 262 BC, the Romans had secured the eastern half of Sicily. Their new fleet of 100 quinqueremes and twenty triremes[15] was also ready but most of the crews lacked experience and were vulnerable to the superior speed and ramming ability their enemy. The Romans' best weapon was their superb infantry and to enable then to board an enemy, they contrived a bridge or *corvus* (raven, see p. 31) to lock the ships together and allow the infantry to cross in good order.[16] In support of this, they built towers on the ships' decks and to the normal complement of forty marines for a quinquereme, they embarked a century of legionaries, approximately eighty men, to give them overwhelming numbers.[17]

The fleet sailed for Sicily in 260 BC and while it was working up, a squadron of seventeen ships

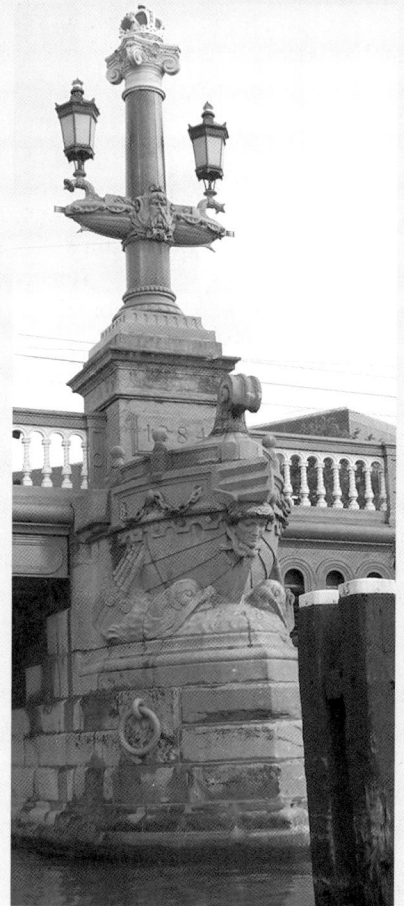

The prow of an ancient warship remains a powerful symbol, especially for monumental edifices. Examples can be as varied as a First World War memorial in Reggio, Italy (upper left), the memorial to the loss of the Maine, in New York, USA (upper right), an admiralty column in St Petersburg, Russia (above), and a bridge in Amsterdam, Holland (right). Although interpreted with varied degrees of accuracy, the intention is unmistakable. (*Author's photographs*)

sailed to the Lipari Islands as a result of a (false) report that the Carthaginians had vacated. It was a trap and all seventeen were captured by the sudden descent of the Punic fleet. Another Punic force of fifty ships also had made a sweep and blundered into the advancing Roman main fleet coming down the Italian coast; the Punic ships were mauled but escaped and the Romans went on to Messina, before sailing to look for their enemy in earnest.

The fleets, 130 Punic ships and 143 Roman, mostly quinqueremes, met off Milazzo. The Romans closed with the Carthaginian ships, dropped the *corvus* and boarded. It was a massive victory and broke the centuries-old Carthaginian domination of the seas, dealing them a blow to morale from which they never recovered.

In the following year (259 BC) the Romans took Corsica and most of Sardinia, where they again beat the Punic fleet twice at Sulci (258 BC). The war in Sicily became bogged down and in 257 BC, the Roman fleet took Malta (Melita) but on their return to Sicily, ran into the Punic fleet off Tindari (Tyndaris). After losing nine out of the ten ships of their advance group, the Roman main fleet then came into action and captured eighteen enemy ships. To circumvent the stalemate on Sicily, the Romans mounted an invasion of the Punic homeland in 256 BC. The invasion fleet sailed westward along the south coast of Sicily; the Punic fleet waited for it behind the headland of Licata (Ecnomus). In the ensuing battle the Romans inflicted yet another crushing defeat on the Carthaginian fleet. The subsequent landings in Africa were initially successful and the main fleet withdrew to Sicily for the winter. The invasion force was later badly beaten and fell back on and fortified Nabeul, Tunisia (Clupea) and so the next spring (255 BC) the fleet, some 200 strong plus transports, was sent to evacuate them. Hearing that the Punic fleet, also of some 200 ships, was off Cape Bon (Hermaeum) the Roman admiral called for the forty ships still at Nabeul to join him. Their enemy was undertrained and with poor morale. The Roman ships outmanoeuvred the Punic fleet and drove it back on to the shore, destroying or capturing no less than 114 of their ships.

The Roman fleet successfully evacuated the army from Nabeul and with its captive warships and men set out for Sicily. There were between 300 and 400 ships, heavily laden, but between Malta and Sicily they ran into a great storm. It was a disaster: only eighty of the warships and some transports limped into port. The rest were lost and with them the best part of 100,000 men perished, the greatest loss of life in a single shipwrecking in the history of seafaring.

A year later (254 BC) the Romans managed to commission a fleet of 170 warships but the loss of veteran crews meant that their replacements had to perform relatively low-level operations while they gained experience. The Carthaginians were, however, tied down quelling revolts in their African territories and unable to take advantage of the Romans' comparative weakness at sea. The new fleet was involved in the capture of Palermo (Panormus) in 254 BC and raided the African coast in the following year. On the return journey, it too ran into a severe storm off Cape Palinuro (Palinurus) losing twenty-seven ships to the rocky coastline.[18]

By 249 BC, with Punic forces isolated in Marsala (Lilybaeum) and Trapani (Drepanum) on western Sicily, the Romans had rebuilt their fleet up to 240 ships. The consul, Publius Clodius Pulcher was sent with 123 ships to Marsala, where the other consul, Pullus with 120 ships, transports, supplies and reinforcements was to join him. Pulcher knew that the Carthaginians had 100 ships at Trapani and that a second Punic fleet of seventy ships was on its way, resolved to attack them without waiting for his colleague. The Roman fleet was outmanoeuvred and totally defeated, suffering their first naval defeat of the war. Upon Pulcher's returning to Marsala, the survivors were attacked by the second Punic fleet, losing another nine ships.

The Carthaginians next sent a fleet of 120 ships to find Pullus' fleet, now reduced to ninety ships. Pullus managed to send his transports to safety but was then attacked by the Punic fleet which forced the Romans back towards the rocky shore south of Camarina. Before the action could be fully joined, the Punic fleet broke off, running to the east, towards Cape Correnti (Pachynus).

They had foreseen the oncoming of another of the violent storms to which the area is prone, leaving the hapless Romans to be driven on to the rocks and wrecked; only two ships escaped. Of the 240 ships at the start of the year, twenty were left. The war stagnated again, the Carthaginians being committed to their African troubles and unable to take advantage of the situation; they had, in fact, to decommission many ships to release manpower for use ashore. The Romans gradually, yet again, rebuilt their strength and undertook raids on the African coast.

By 242 BC the once more rebuilt Roman fleet was ready and after escorting a supply convoy to western Sicily, went looking for the Carthaginian fleet. The Romans had abandoned the *corvus* for a lighter, less top-heavy alternative and were using a much better designed type of ship. In March 241 BC a Punic fleet of between 120 and 150 ships sailed with supplies for their very hard-pressed garrisons on Sicily. The Roman fleet had been training hard and was in the peak of condition, whereas their opponents were undermanned, having planned on being able to unload the supplies and augment their crews from the garrisons. The Romans caught them first, off the Egadi Islands (Aegades) on 10 March, ramming and sinking fifty of their ships and capturing seventy more, the few remaining fleeing back to Africa.[19] Without the supplies, the garrisons had to surrender and for Carthage it was the final blow: she sued for peace.[20]

The war had cost Carthage about 450 warships, all lost in combat to the Romans. Roman losses amounted to just under 600 ships, only 200 of which were to enemy action, the rest being lost to storms. Rome ended with the stronger navy and gained Sicily and, a little later, Sardinia. Carthage, however, was otherwise intact and still had what was left of her navy, without restrictions on rebuilding.

As the victor, Rome was free to expand and her navy was instrumental in extending Roman power to Liguria in the west and to the northern Adriatic in the east. Campaigns on the Dalmatian coast (Illyria) and southern Adriatic suppressed piracy and established a Roman territory in present-day Albania. A recovering Carthage had however been expanding its power once more and it would not be long before further conflict occurred.

In the resumption of the struggle that was the Second Punic War (218–201 BC) the Roman navy's command of the sea and the comparative lack of activity on the part of the potentially still powerful Carthaginian navy, meant that, for all of his tactical genius and the crushing victories that he attained against Roman armies, Hannibal could not win. An aggressive policy by the Roman navy meant that the Romans could always hold the initiative and dictate the course of the war. This war also saw the Romans drawn into their first operations in Greece (the First Macedonian War, 215–205 BC), in which it was Roman naval operations that prevented a joining of Macedonian and Hannibal's forces that could have proved fatal for Rome.

At the outbreak of the war, the Romans had planned for a fleet of sixty warships to support the landing of an army in Punic Spain, while another, of 160 warships would attack the enemy homeland in Africa.[21] The Carthaginians had greatly extended their territory in Spain and it was their siege of Sagunto (Saguntum) that was the trigger for the war. It was obvious that, with Carthage relying to a large extent on Spain for resources, it would be a major focus of the war. Roman sea power and Punic unwillingness to challenge it forced Hannibal to make his famous march to Italy, instead of simply sailing there. Behind him, the Romans had, in 218 BC, established a naval base at Tarragona (Tarraco) cutting his connection with Spain and taking the war and the initiative to the enemy there.

There was a Punic fleet of thirty-two quinqueremes and three triremes in Spain and in 217 BC, the Romans, with thirty-five quinqueremes caught the Punic fleet anchored and pulled into the shore near the mouth of the River Ebro (Iberus) with many of the crews ashore. The Romans attacked, taking two and swamping another four ships. The Punic crews managed to scramble aboard but could not break out and ran the ships ashore, fleeing to the safety of their nearby army. The Romans pressed their attack and took a total of twenty-five enemy ships, breaking Punic naval power in Spain, which would never again challenge them there. A Punic fleet sailed to Sardinia and Corsica in 217 BC and managed to capture a Roman supply convoy but fled at the approach of the Roman fleet. The latter then raided the African coast, repeating the act in the following year, but with little success.

Meanwhile, Hannibal had inflicted serious defeats on the Romans, culminating in the disaster of Cannae in 216 BC. The losses meant that men had to be drafted from the fleets to augment the army and fifty ships laid up. Even so, the Roman navy continued to dominate and Punic squadrons probing towards Sicily were driven off.

In a dramatic coup, the Roman navy intercepted five ships off southern Italy which carried envoys and the terms of a treaty between Hannibal (then in Italy) and Philip V (reigned 221–179 BC), the expansionist king of Macedon. The Romans increased their Adriatic fleet from twenty-five to fifty ships and assembled an alliance of Greek states nervous of Philip's ambitions, to confront him (215 BC). Having isolated the Macedonian War, a new problem for Rome arose in Sicily when their ally there, Syracuse, joined Carthage. To meet this added threat and neutralise the Syracusan fleet, the Roman fleet in Sicily was increased to 100 ships and an army sent to besiege the city.

In 214 BC Philip attempted a naval foray into the Adriatic but a swift response by the Roman fleet there forced his larger number of smaller ships back up a river until they were helpless and had to be burnt. An extra 100 warships were ordered to be built by the Romans in that year[22] and a huge naval and land assault was mounted against the formidable defences of Syracuse. The assault failed, despite meticulous preparation, foiled by the artillery and other machines employed by the defenders, guided by Archimedes.

Hannibal meanwhile, moved into Taranto but was denied the safe harbour as the citadel of the city was held by the Roman garrison, which continued to hold out for four years, supplied only by sea. A Carthaginian fleet of over 100 ships brought a supply convoy towards Sicily but sailed away upon the appearance of the Roman fleet. With no hope of relief, Syracuse fell in 211 BC. A raid by the Roman fleet on Utica captured nearly 130 laden transports, ironically, used by the Romans to relieve the fallen Syracuse.

In the east the fleet operated with its allies in the Aegean, retiring to Corfu and the Gulf of Corinth for the winters. In 210 BC the Tarentines managed to get about twenty of their ships to sea and intercepted a mixed twenty-ship Roman squadron bringing supplies for their garrison and defeated them. The following year (209 BC) in a surprise attack, Roman forces in Spain stormed the Punic capital of Cartagena (Carthago Nova) backed by eighty ships which assaulted the seaward walls of the city. In the next few years, Roman fleets continued to raid the enemy's African coastlines and to operate in Dalmatia and the Aegean, as well as dominating the coast of Spain. Hannibal moved out, so the Romans reoccupied Taranto, further isolating him. On one raid in 208 BC the Roman fleet came upon a Punic fleet of eighty-three ships and in the only the second fleet battle of the war, defeated it, taking eighteen ships.

By 206 BC the Romans were driving the last Punic forces out of Spain and advancing on Cadiz (Gades). A Roman squadron sailed past Gibraltar, into the Atlantic for the first time, finding and ramming and disabling three ships of a Punic squadron of eight, running for Africa. The final act of the war was the Roman invasion of the Carthaginian homeland in 204 BC, escorted and kept supplied by the fleet. There was one final foray by the Punic navy against the Roman fleet which failed. Although he managed to avoid the Roman patrols and return to Africa, Hannibal was defeated at the battle of Zama in 202 BC, ending the war. The Carthaginian fleet, which had failed to seriously challenge Roman naval power during the war, with the exception of ten triremes, was towed out to sea and burnt.

Gaius Iulius Caesar (100–44 BC). Staff officer to the Dictator Sulla in 83 BC, he rose to be Governor of Spain by 61 BC. His conquests there and skilful political manoeuvring enabled him to form the First Triumvirate with Pompeius and Crassus (between 60 and 53 BC). He conquered Gaul for the Empire between 58 and 50 BC and after the death of Crassus, led one of the factions competing for power against the other, led by Pompeius. Civil war between them led to Caesar's overall victory and his becoming dictator for life in 45 BC. His reforms and intended campaign against Parthia were stopped by his assassination in 44 BC. An original bust; Naples Archaeological Museum. (*Author's photograph*)

attack by Caesar's troops caught most of their crews ashore and set enough of the fleet alight to break the blockade and send for help. When it arrived, the relieving force was attacked off the harbour by the remaining Egyptian ships, which were beaten off. More reinforcements brought Caesar's fleet up to twenty-five Roman ships (five quinqueremes, ten quadriremes, ten triremes and others) and nine Rhodian ships. The Egyptians had also concentrated their ships and had five quinqueremes, twenty-two quadriremes and some smaller ships.

Avoiding a fireship attack, Caesar's fleet sailed from the eastern to the western harbour to attack the Egyptian fleet anchored there. After a hard fight, the Romans and Rhodians disabled and captured several of the enemy ships, the remainder escaping through a canal into Lake Mareotis. They made one more sally to intercept more Roman reinforcements but Caesar's fleet again beat them off. Ptolemy's army was defeated and Ptolemy killed; Caesar installed Cleopatra as the sole ruler, backed by a Roman garrison and left for Rome.

One of Pompeius' admirals was still resisting in the Adriatic but in spring 47 BC, Caesar's commander at Brindisi (Brundisium), with a scratch fleet but superior soldiery, beat him. The remaining forces of Pompeius in Africa had a fleet of fifty-five ships when Caesar attacked them, with his fleet of forty ships. After

some skirmishing, they drove Pompeius' fleet, whose morale was low, into harbour at Sousse (Hadrumetum) and held them there until the end of the campaign. The forces of Pompeius were destroyed at the battle of Munda in March 45 BC, his sons Gnaeus and Sextus escaping to Spain with some ships.[31] Gnaeus, with thirty ships was caught by Caesar's fleet near Gibraltar and killed in the ensuing battle but Sextus got to Spain and started to raise troops there.

Caesar was murdered in 44 BC, setting off the next round of civil wars with the plotters Cassius and Brutus, Caesar's nephew and heir Octavian, his former general Antonius and Pompeius' younger son, Sextus, the principal players. The Senate sent Cassius east with a fleet, where he destroyed, firstly Caesar's eastern fleet and then took Rhodes, ending its naval power.[32] The Senate made Sextus commander of the navy and he made Sicily his power base, securing it with a fleet of over 130 ships. Cassius and Brutus marshalled their forces in Greece; Octavian and Antonius joined forces but were blockaded in Brindisi by a fleet of sixty ships, later increased to 110. They managed nevertheless to cross the Adriatic with their army when the enemy fleet was known to have moved away and six weeks later, defeated Cassius and Brutus at Phillipi (42 BC). Their admiral, not knowing of the defeat, later intercepted and took a supply convoy and its escort.

Pompeius the Great (Gnaeus Pompeius Magnus) (106–48 BC). Rose to prominence as a military commander under the dictator Sulla, who granted him the *cognomen* Magnus. Given overall command in 67 BC, he led Rome's forces to sweep piracy from the seas, going on to defeat Mithridates of Pontus and march to Armenia. He annexed Syria and settled affairs in the Levant. He joined the First Triumvirate with Caesar and Crassus that effectively ruled between 60 and 53 BC, when Crassus was killed. Loss of the balance of power provided by the Triumvirate led to civil war between him and Caesar, in which he commanded the allegiance of most of the Roman fleets. The war was ended with his defeat by Caesar at the battle of Pharsalus and Pompeius' flight to Egypt, where he was ignominiously murdered. Louvre. (*Author's photograph*)

By 40 BC Octavian held Italy and the west and Antonius, who had inherited the eastern fleet, held the east. Sextus, now with over 200 ships based on Sicily, was interfering with maritime trade and grain shipments to Rome. The inevitable war between Octavian and Sextus came in 38 BC. Octavian led his army and fleet of ninety ships, with newly raised crews, south to rendezvous at Reggio di Calabria (Rhegium) with another squadron coming from the Adriatic. Sextus' crews were experienced and, leaving ships to guard the Strait of Messina, he sailed with about ninety ships, catching Octavian's fleet, which had already suffered damage in a storm. In a battle off Cumae Octavian's fleet was mauled but drove Sextus off. A second very similar fight occurred shortly afterwards, when Sextus' returning fleet again engaged that of Octavian. Although he had to disengage on the approach of the Adriatic squadron, Sextus had destroyed half of Octavian's fleet.

To resume the offensive, Octavian appointed his friend Marcus Vipsanius Agrippa to command the fleet. The fleet was enlarged, trained and new weapons introduced. In 36 BC Octavian led an army to the Straits, backed by a fleet of 100 ships, 'lent' by Antonius; another force, with seventy ships from Africa, landed on western Sicily and Agrippa advanced from the north with the main fleet. Sextus' fleet attacked a supply convoy from Africa, sinking half and scattering the rest. A raid on Naples was unsuccessful and he was worsted in an encounter with Agrippa's ships, losing thirty ships to Agrippa's loss of only five. Octavian had meanwhile started to land near Taormina (Tauromenium) but Sextus' fleet appeared, taking or sinking about sixty of his transports and driving off Antonius' fleet with ease. The net was closing on Sextus, however, and

Despite the insecure nature and brevity of his reign (42–36 BC) Sextus Pompeius, the younger son of Pompeius, is still commemorated in Sicily, if nothing else, at least by this street name in Taormina. (*Author's photograph*)

After his victory at Actium, Octavian built Nicopolis, 'victory city', on the site of his former encampment, moving the population of the nearby Greek town of Kassope to inhabit it. On the hilltop where his headquarters had been, he had a huge victory monument erected and around the stonework base of it the rams taken from captured ships were mounted. The base of the mound, right, can be seen; it had a victory shrine within a colonnade on top; on the retaining wall at the base of the monument, below, the mounting sockets for the rams can be seen. (*Author's photographs*)

finally his massed fleet met that of Agrippa off Naulochus in 36 BC, resulting in the total defeat of Sextus' fleet. Sextus fled to Antonius, who had him executed.

The final struggle, between Octavian and Antonius came in 33 BC. Antonius and Cleopatra moved, with the eastern and Egyptian fleets, into western Greece and by late 33 BC were encamped at Actium on the Gulf of Ambracia. Octavian crossed with his army, covered by his fleet, again commanded by Agrippa. In a masterful maritime campaign, Agrippa choked Antonius' supply routes, which resulted in the battle of Actium in September 31 BC. With the death of Antonius and Cleopatra and the defection of the Egyptian navy to Octavian in the following year, the Roman navy became the only one in the Mediterranean Sea.

Imperial fleets

At the end of the civil wars, Octavian had nearly 700 warships of all types, far more than would ever be needed; the numbers were reduced by laying up, scrapping or simply burning surplus ships. Octavian adopted the title Augustus in 27 BC and with Agrippa reorganised naval forces to form the imperial fleets. There followed a period, unequalled before or since, of approximately three centuries during which that navy was to exercise unchallenged supremacy over the whole Mediterranean basin, extending its influence in time to encompass the Black, Red, North and Irish Seas, the English Channel and north-west Atlantic seaboards, as well as the great river frontiers of the Empire, the Rhine and Danube (Danubius). These forces reigned supreme until in the late second century AD, when the first barbarian raiders from across the North Sea were noted, but they were not then enough to provide any serious challenge.

The fleets, apart from policing the seas, were also engaged in offensive operations, starting in 25 BC, with the transfer of a fleet of eighty warships to the Red Sea to support 130 transports and a military expedition to Sabaea (probably modern Yemen). In 17 BC Agrippa commanded a fleet which landed marines in the Crimea to resolve a dynastic problem in the Bosporan kingdom, a client state. Between 17 and 15 BC Augustus advanced the Roman border to the line of the Danube, forming river fleets for it. Fleets operated along the Dutch and German coasts as well as into the rivers of Germany after 12 BC in support of Augustus' short-lived province there. With peace and security in the Mediterranean came an increase in maritime trade, reaching in the first two centuries AD, a level not surpassed until the nineteenth century. Merchant ships grew in size as well as in numbers and regular sailing schedules for cargo and passengers became the norm.

The next major challenge for the navy was in AD 43 with Claudius' invasion of Britain; for the invasion army of four legions plus auxiliaries

(about 40,000 men in total) a fleet of over 300 warships and transports was assembled. Apart from ferrying the army, it would be the sole means of supply, the invasion forces not planning to live off the land at all. The fleet was also to accompany the army's advance along the south coast and up the Thames (Tamesis) guarding the flanks, outflanking any opposition and supplying the troops. These formations, which would later become the *Classis Britannica*, continued the process of keeping the flanks of the Roman advance secure throughout the invasion, with fleets on both east and west coasts of Britain.

The navy supported Claudius' annexations of Mauretania (the coastal regions of Algeria and Morocco) in AD 41 and 42, Lycia (south-west Turkey) in AD 43 and Thrace in AD 46, giving Rome control of the whole Mediterranean seaboard. The extension of Roman naval power in the Black Sea was completed by the founding of a major naval base at Chersonesus (near Sevastopol, Crimea) in AD 45 and the annexation of the kingdom of Pontus (on the north Turkish coast) in AD 64. By this time also, Roman warships were operating in the North and Irish Seas. As a portent of things to come, however, in AD 41 tribesmen from the north German coast, using small open boats, raided the coast of Belgium before being driven off. They repeated the exercise in AD 47, burning an auxiliary fort at the mouth of the Rhine but were caught by ships of the *Classis Germanica* and destroyed.[33]

In AD 59 Nero (emperor AD 54–68) conspired with his prefect of the Misene fleet, Anicetus, to murder his domineering mother, Agrippina. The first attempt using a collapsible boat failed when she proved to be a strong swimmer. Anicetus then had his sailors simply batter her to death. Anticipating unrest to come, Nero formed two legions, I and II Adiutrix, from loyal marines of the Misene and Ravenna fleets respectively, but committed suicide in AD 68. This triggered the events of the so-called Year of the Four Emperors, which involved the Italian and Rhine fleets. Nero's 'navy' legions surrendered to his successor Galba but suffered casualties at the hands of his men. By January 69 Galba was dead, succeeded by Otho, who regained the loyalty of the Misene fleet and naval legions. He moved north to oppose the advance of Vitellius, proclaimed emperor by the Rhine fleet and legions. Otho sent the Misene fleet to secure southern Gaul for him but with poor morale and lax discipline, the fleet took to raiding and plunder. After two fierce battles against Vitellius' forces, the fleet withdrew. Otho, relying heavily on naval personnel, was defeated at the battle of Cremona in April AD 69, leaving Vitellius emperor.

In the east, in July AD 69, Vespasian was proclaimed emperor and an army supporting his claim, proceeded to Italy, supported by the *Classis Pontica*, gathering further support as it advanced. Vitally the Ravenna fleet declared for Vespasian, securing his forces' flank and lines of supply and tipping the balance firmly in his favour. Vitellius was defeated at the second battle of Cremona in October AD 69, but with his remaining supporters, continued the struggle until killed that December. For the Misene fleet it was a time of confused loyalties, following their ignominious support of Otho. Factions pulled the fleet between Vitellius and Vespasian until the latter's cause prevailed.

In the meantime, there arose a serious revolt led by Iulius Civilis, a commander of Batavian auxiliary troops (from north-eastern Belgium), who seized some of the German fleet's ships and occupied the mouth of the Rhine. With the ships, augmented by the boats of German tribesmen from east of the Rhine, he pushed upriver against garrisons weakened by troops having left to support Vitellius. Defections by other auxiliary crews and troops, forced the remains of the fleet and legionaries to withdraw to Bonn (Bonna) where they were beaten by the rebels, again joined by Germans from east of the river. Xanten held out and although the fleet sallied successfully against the rebels, it had insufficient strength to be decisive. The Romans fell back on Trier (Augusta Treverorum) with their fleet at Koblenz (Confluentes).

With the end of the civil war, forces could be sent to restore order: the *Classis Britannica* landed troops in Belgium, pushing the rebels back but in the process, losing a lot of their ships to an attack by tribesmen from what is now Friesland.[34] The Rhine fleet remained in poor condition and lost some moored ships to German raiders.[35] With increasing Roman success, it recovered and sailed against Civilis' fleet. Although coming close to each other and exchanging missiles, conditions did not permit an engagement. Civilis surrendered shortly afterwards.[36]

The navy resumed its normal 'peacetime' duties and operations. The *Classis Pontica* secured the south-east Black Sea area and in AD 79 the Misene fleet assisted in rescue attempts during the great eruption of Vesuvius that buried Pompeii and the surrounding area. Ships of the *Classis Britannica* circumnavigated Britain in AD 80 and reconnoitred Scotland and Ireland two years later. The *Classis Germanica* was in action repelling Germanic incursions across the Rhine and the *Classis Moesica* had helped in stopping a Dacian invasion across the Danube in the late eighties AD.

The beginning of the second century AD was marked by the Dacian Wars (AD 101–106) of Trajan (emperor AD 98–117). The Danube fleets were augmented by ships seconded from other, seagoing fleets. From the frequency with which and the number of ships shown on Trajan's Column in Rome, which recounts his campaigns, it is clear that they relied principally upon ships operating on the Danube and up its tributaries on both banks for their supplies. Trajan also restored the Nile–Red Sea canal and sent a fleet into the Red Sea to support his annexation of the Nabatean kingdom in AD 106. The *Classis Alexandrina* expanded its remit to include occasional patrols in the Red Sea, although it does not appear that any permanent naval presence was stationed there. This fleet also took over policing of the River Nile from the former Ptolemaic river force.

The reign of Hadrian (emperor AD 117–138) was marked by comparatively minor activity, the ferrying of troops by the German fleet to assist in building his wall in Britain in AD 122; an expedition to bolster the Bosporan kingdom; assisting in suppressing a revolt in Judaea. Comparative peace continued through the reign of Antoninus Pius (emperor AD 138–161). A Parthian offensive through Armenia and Syria in AD 161, led to a realignment of the eastern fleets, with the *Classis Pontica* moving west to Cyzicus and the *Classis Syriaca* being greatly strengthened.

In the latter half of the second century AD the Danube fleets were in constant action dealing with the irruptions of barbarians across the upper and middle Danube and supporting the wars of Marcus Aurelius (emperor 161–180) against them. Between AD 170 and 171 the Misene fleet operated off the Atlantic coast of Mauretania (Morocco), helping to put down tribal unrest. The end of the century saw the British fleet join a punitive expedition to Scotland (AD 185), the start of barbarian sea raiding along the northern shores of the empire and the building of forts to oppose them. It also saw another civil war from which Septimius Severus (emperor AD 192–211) emerged as emperor. The murder of his son and successor, Caracalla, in AD 217 initiated a period of instability in government with no less than twenty-two emperors and many usurpers. This period of turmoil, when barbarian pressure from beyond the borders, allied to a rejuvenated and aggressive Persian empire and coupled with internal instability and dissention, started to threaten the integrity of the empire and provided the first serious naval challenge for nearly three centuries. Further this was at a time when, once more, the Romans, lulled into a complacency by long secure and peaceful sea lanes and with a navy that had for too long been resting on its laurels, had it seems, allowed that navy, or at least large parts of it, to deteriorate. An increasing incidence of raids from across the North Sea was coupled with ventures by Goths from North of the Black Sea across that sea against Roman shores, even extending their activities into the Aegean; once again, the neglect had also permitted a resumption of that old scourge of the seaways, piracy.

Roman control of the Black Sea weakened and by mid-century Goths had occupied its northern littoral and were raiding by boat. In AD 259, a Goth fleet of as many as 500 boats, penetrated into the Aegean. They repeated the feat in AD 268 but were intercepted by Roman naval forces and twice defeated. Yet another

Goth fleet attacked Thessaloniki in AD 269 but having had to abandon their siege of the city; their fleet was destroyed in a series of running battles, by the *classes Alexandrina* and *Syriaca*. In the north a system of extensive coastal defence works, allied to naval squadrons was formed along the coasts of Britain and France, to oppose increasing barbarian seaborne activity and ability. Pressure on the river frontiers taxed those fleets as the empire tottered. The navy's formations struggled to maintain their presence in the face of declining resources and increasing enemy activity.

By the time of the accession of Diocletian (emperor AD 285–305) and the re-establishment of stability at the very end of the third century AD, a very different navy remained; the grand Imperial fleets were no more and the service was reorganised into a number of smaller squadrons, each allotted to a particular area. The formula was successful and for another century the Imperial navy, in all of its various dispositions, again kept the peace in the Mediterranean and minimised barbarian activity in the North and Black Seas although, with the probable exception of the 'British' fleet in the north, the service never again reached the size and strength that it had previously enjoyed. Without the large and strong central Italian fleet organisations to give it a strong separate identity, the scattered squadrons of the navy seem to have lost much of their former independent nature as each squadron was placed under the command of and seen as subordinate to the local senior military commander.

In AD 286 a local commander, Carausius, was appointed to command the *Classis Britannica*. He built up the fleet, improved efficiency and successfully attacked barbarian sea raiders. Noticeably the loot recovered was not always returned and the emperor Maximius (emperor AD 286–305) sentenced Carausius to death, he promptly declared himself emperor and withdrew his fleet and forces to Britain, retaining Boulogne (Gesoriacum) and Rouen (Rotomagus) in Gaul. An attempt by Maximius to end the secession in AD 288 failed when his scratch fleet was defeated by the veteran

Classis Britannica. In AD 290 a treaty was made acknowledging Carausius' position. He strengthened the economy in Britain, the fleet and further extended the shore defence system.

In AD 293 Carausius lost his holdings in Gaul, whereupon he was assassinated and succeeded by Allectus. In AD 296 the Caesar Constantius with a larger and much improved fleet, evaded Allectus' fleet and landed troops, defeated him recovering Britain for the empire. Constantius' ships then ranged widely, even carrying out a punitive raid on the Orkneys (Orcades).[37] The *Classis Britannica* was brought back to duty and the shore defence system further extended. The result was a notable decline in barbarian activity for nearly half a century.

The role played by the river fleets on the Rhine, Danube and their tributaries when they were not engaged supporting military campaigns into barbarian lands was unrelenting. The crews had to be constantly vigilant. Not for them the luxury of spotting a ship on the horizon, and thus probably hours away; on the rivers enemy craft could emerge from a bank and be in contact in minutes. If they sailed closer to the barbarian bank, a shower of arrows could come suddenly from behind the trees and bushes; they could look upriver and see a swarm of raider's boats cross long before they could turn and row against the current towards them. The borders had never been absolute barriers but the increasing numbers and frequency of barbarian activity severely taxed the river fleets, ultimately they would be overwhelmed.

Civil war returned in the fourth century AD and with it the first great fleet battle since Actium, between the fleets of Licinius (emperor of the East, AD 308–324) and Constantine (emperor in the West, AD 307–324; sole emperor AD 324–337) in the Dardanelles (Hellespont) in AD 323. Licinius is said to have had a fleet of 200 or more triremes, with which he sought to block the Dardanelles against Constantine's fleet of eighty or more 'triaconters'. This was an archaic term which, from the result, appears to have been used to describe a new, more powerful type of warship, the result being that Constantine's fleet won a battle after which the remaining enemy ships withdrew from the war.

The empire remained reasonably strong for the rest of the fourth century AD despite the continual barbarian pressure and incursions across the river frontiers and North Sea. A change of strategic emphasis from defending the borders to having mobile forces to destroy invaders meant that more of the fighting was done on Roman territory, to its detriment, while the river forces were only expected to deal with minor incursions and only delay larger ones, their morale being affected accordingly.[38]

There were still successes, such as in the winter of AD 357 when a band of Franks seized an old Roman river fort; the future emperor Julian (emperor AD 361–363) surrounded it and the *lusoriae* of the river fleet patrolled to keep the ice broken and prevent escape; the Franks surrendered.[39] In AD 386 a huge number of Goths tried to force a crossing of the Danube and migrate to Roman territory, using a vast number of boats and rafts. The Romans concentrated their Danube squadrons, augmenting them with seagoing warships and totally destroyed the invasion.[40]

By the end of the century, control of the North Sea had been lost and in the winter of AD 406 the Rhine froze solid, immobilizing the fleet and permitting the mass migration of barbarians into Gaul. Twenty years later the Vandals had obtained ships in Spain and crossed to North Africa.[41] In AD 429 they took Carthage and set about creating a navy, the first non-Roman navy in the Mediterranean for four centuries. Large tracts of the western empire became barbarian kingdoms and the Rhine and British fleets disappeared. The Danube and eastern fleets remained but in the west, growing Vandal power gained them Sicily, Sardinia and Corsica. A fleet sent against them was defeated in AD 440 and they destroyed a western fleet fitting out in Spanish ports in AD 457. In that same year, the Vandal fleet was defeated at Ostia.[42]

The final act for the Roman navy was in AD 467 when, with an eastern fleet, it drove the Vandals from Sardinia. There followed an attempt by the joint fleet to convey an army to recover Africa. The army was landed near Cape

The Iron Gates, the Great Kazan, looking downstream. Before the river was dammed downstream, the gorge was much deeper with rapids and not navigable. As such it was the dividing point between the remit of the *Classis Pannonica* (upstream from here) and the *Classis Moesica* (downstream to the sea). For his Dacian Wars, AD 101–106, Trajan built a military road into the cliff faces on the right bank and a canal with locks, 2.5 miles (4 km) long, along the base of the cliffs on the right to enable river traffic; both are now below water level. (*Author's photograph*)

Bon but delayed their attack. Five days later, with a favourable wind, the Vandals launched an attack with fireships against the tightly packed Roman ships, anchored against a lee shore, following it up with a ram attack. They destroyed half of the Roman fleet, thwarting the invasion.[43] It was to be the last action by a Roman fleet from the western half of the empire that would end with the abdication of its last emperor (Romulus Augustulus) in AD 476. The Eastern part of that Empire would survive for another thousand years as a body politic that for convenience, we refer to as Byzantine and would continue to operate a navy, but which is as such, part of another story.

CHRONOLOGY OF ROMAN NAVAL AFFAIRS

753 BC Traditional date of the founding of Rome

394 Earliest record of a Roman warship

338 First recorded Roman naval action; victory off Anzio

326 Samnite naval raids on Roman coast; Roman experiments with naval forces

312 First Roman overseas possession, occupation of the Pontine Islands

311 Appointment of *duovirii navales*

310 Roman naval attack on Pompeii and area

282 Roman warships defeated by Tarentines in Gulf of Taranto

267 Four *praefecti* of the fleet appointed

264–241 First Punic War. Naval landings on Sicily

260 Battle of Mylae, Romans defeat Punic fleet

258 Roman fleet defeats Punic fleet at Sulci, Sardinia

257 Roman fleet raids Malta; defeats Punic fleet at Tyndaris, Sicily

256 Roman victory at battle of Ecnomus, Sicily. Roman army landed in Africa; Punic fleet defeated off Hermaeum, Tunisia

255 Roman fleet evacuated army from Africa, nearly all lost in a storm

254 New fleet aids capture of Palermo, Sicily

253 Roman fleet raids African coast, wrecked by storm

250 Fleet operates in western Sicily

249 Battle of Drepanum, only Roman naval defeat of the war; enemy attacks and storm wrecks remainder of Roman fleet

242 New Roman fleet, paid for by public subscription, sails for Sicily

241 Battle of Aegades Islands, Roman fleet destroys Punic fleet and ends war

238 Corsica and Sardinia annexed

229–228 Fleet operations in the Adriatic. Roman protectorate in Illyria

220 Fleet ferries army across Adriatic and clears it of pirates

218–201 Second Punic War. Navy base at Tarragona in Spain

217 Roman fleet defeats Punic Fleet at battle of the Ebro

216 Naval raids on African coast

214 Assault by Roman fleet on Syracuse defeated. Macedonian fleet defeated in Adriatic

211 Fleet attack on Utica

210 Roman fleet in the Aegean. Romans beaten at sea by Tarentines. Raids on African coast

209 Navy and army jointly take Cartagena. Eastern fleet in Gulf of Corinth. Sicilian fleet raids Africa

208 Raids on African coast, defeat of Punic squadron there

207 More raids on Africa. Punic fleet defeated again. Eastern fleet in the Aegean

206 Battle in Strait of Gibraltar, first Roman ships in the Atlantic

205	Raids on African coast
204	Navy escorts invasion army to Africa
203	Punic attack on fleet in Africa fails
202	End of war, Punic fleet surrendered
200	A fleet sent to the Aegean to join local allies
199	Roman and allied fleet secure Aegean
197	Macedonian fleet surrendered
195	Roman and allied operations against Sparta, whose fleet is surrendered
190	Roman and allied fleet defeat Seleucid fleet at battle of Myonnesus
189	Seleucid fleet surrendered
188	Anti piracy operations
178	Operations in northern Adriatic
172	Fleet returns to Aegean
154	Naval operations in the Atlantic on coast of Portugal
149–146	Third Punic War. Fleet transports army to Africa and blockades Carthage
147	Last foray of Punic navy, which is beaten
139	Navy ships reach Atlantic end of the Pyrenees
122	Fleet captures Balearic Islands. Period of naval decline
86	Sulla has to gather a scratch fleet to campaign against Mithridates
84	Ad hoc eastern fleet disbanded
79	Anti piracy sweep off Cilicia
78	More anti piracy operations
77	Pirate fleet defeated off Lycia
76	More anti piracy operations
73	Fleet sent east but beaten by Mithridates in Bosporus. Marcus Antonius operates against pirates
72	Marcus Antonius beaten by pirates in Crete
67	Navy in wretched state. Piracy rages. Pompeius renovates navy; war against the pirates destroys them
66	Roman Black Sea and Aegean squadrons formed
58	Navy assists in annexation of Cyprus
56	Caesar's campaign in Brittany. Roman fleet defeats Gauls
55	Caesar's first expedition to Britain
54	Caesar's second expedition to Britain
50	Naval forces stationed in English Channel
49	Civil war. Pompeius' fleet wins Adriatic. Naval battles off Marseilles during siege
48	Caesar's siege broken by Pompeius' fleet
47	Egyptian fleet defeated by Caesar's
43	Sextus Pompeius commands navy. Octavian forms his own fleet
42	Sextus controls shipping to Rome

CHRONOLOGY OF ROMAN NAVAL AFFAIRS *continued*

38	Sextus defeats Octavian's fleet off Cumae
36	Agrippa defeats Sextus off Mylae. Battle of Naulochus ends Sextus' rule
33	Civil war. Eastern and Egyptian fleets join Antonius at Actium
31	Agrippa's naval campaign ends in battle of Actium
30	Egyptian navy surrenders to Octavian
26	Red Sea expedition
25	Reform of navy, founding of imperial navy and fleets
22	Founding of fleet bases at Misenum and Ravenna
17	Fleet operations in Black Sea
15	Advance to Danube, new fleets founded
12	*Classis Germanica* founded. Fleet operates on Dutch Lakes and North Sea
9	Fleet operations on German rivers and North Sea coast and Baltic entrances
AD 6	Judaean navy absorbed
14	Fleet supports campaign in Germany
16	Further operations in Germany
40	Gaius (Caligula) gathers forces on English Channel
43	Invasion of Britain, *Classis Britannica* founded. Invasion depends on navy for supplies
44	Fleet operations in Britain
45	Naval expedition to Bosporan kingdom
46	Temporary fleet formed for the annexation of Thrace
47	British fleet rounds southwest of Britain
52	Operations in South Wales
64	Pontus annexed, *Classis Pontica* formed
65	Nero forms two legions from Italian fleet personnel
69	Civil war. Misene fleet supports Otho, attacks Liguria. Ravenna fleet joins Vespasian.
71	*Classis Britannica* in North Sea
79	Vesuvius erupts, Misene fleet assists in rescue
80	British fleet operates on east and west coasts and circumnavigates Britain and Scotland
81	Navy pay rates increased by Domitian
82	Naval reconnaissance of Scotland and Ireland
83	Navy in combined operations in Scotland
89	Incursions across Rhine defeated by *Classis Germanica*
101	Dacian wars, Danube and Black Sea fleets assist
106	Roman ships taken into Red Sea
127	Navy ships in Red Sea again
161	*Classis Pontica* moves to Cyzicus on the Sea of Marmora
169	Danube fleets active to prevent barbarian incursions. Misene fleet operations off Mauretania
174	Start of barbarian raids on north European coasts

175   Navy enlistment term extended

185   Fleet supports campaign in Scotland

219   Fleet supports an (unsuccessful) usurper

230   Raiding and piracy in North Sea. Start of northern coastal defence system. North Sea coastlines change due to inundation

233   Rhine frontier breached, *Classis Germanica* heavily engaged. Constant action needed in the years following both on Rhine and Danube to resist barbarian pressure

254   Goth sea raids begin in Black Sea

259   Goth fleet breaks through into the Aegean

260   Misene fleet puts down revolt in Africa

268   Barbarian incursions across Danube. Goth sea forces beaten in Aegean

272   Aurelian defeats Goths on the Danube

282   Probus defeats invasion across Danube

283   Danube fleets in constant action in campaigns

286   Carausius commands *Classis Britannica* and secedes. Diocletian reforms army and navy

296   Constantius invades Britain, ends secession. Punitive cruise by fleet

306   Piracy and raiding in Gaul

323   Civil war, battle of the Hellespont. Constantine's forces defeat those of Licinius

328   Campaigns across the Rhine, border and fleet reinforced

332   Similar campaign across the Danube

353   Resumed raiding across river and coastal borders

363   Emperor Julian constructs a fleet on the Euphrates to attack Persia

367   Barbarian conspiracy, invasion of Britain

369   Theodosius restores Britain, strengthens defences

383   Emperor Gratian killed, western fleet supports usurper in Italy

386   Battle on the Danube destroys Goth migration

388   Civil war, battle between rebel and eastern fleets off Sicily, rebels defeated

395   Any control of North Sea lost by now

406   Rhine freezes, fleet immobilised. Mass influx of barbarians into Gaul

429   Vandals obtain ships, cross from Spain to Africa

439   Vandals take Carthage and build a navy

440   Vandals take Sicily, Sardinia and Corsica. Eastern emperor sends a fleet, which they defeat

456   Romans defeat Vandal fleet off Corsica

457   Vandal fleet defeated at Ostia. Roman fleet destroyed by them while fitting out in Spain

467   Joint east and west Roman fleet recovers Sardinia. Fleet attempts invasion of Carthage. Vandal attack destroys fleet. Last action by a western Roman fleet

476   Last emperor of the West leaves office, 'official' end of the Western Empire

PART II THE SHIPS

2 SHIPBUILDING

Construction methods

Marine archaeology has, in the absence of surviving descriptions, yielded detailed information about the methods employed in building Roman ships. The remains of many hundreds of wrecks of seagoing Roman ships have been found in all parts of the Mediterranean,[1] dating from between the third century BC and fifth century AD; none of them, however, can be identified as being warships.[2] A number of river craft have been recovered of which some can be identified as warships. Five such vessels were found at the base of the *Classis Germanica* at Mainz (Moguntiacum) and another two near to Ingolstadt on the Danube. Enough remained for full-size replicas to be built and tested.[3]

In the building of a hull, the earliest Mediterranean boatbuilders formed the 'shell' first and thereafter inserted frames, thwarts, stringers and other hull timbers. The hull planking was joined edge to edge and various methods of fixing them to each other were evolved, such as sewing, using cleats or staples, or nailing. By at least 1350 BC (the date of the earliest wreck so far discovered that uses it[4]) a new method had emerged, namely the use of interlocking mortises and tenons. To build the ship, the shipwrights first set up the keel, stem and stern posts; to this they next attached the garboard strakes, the first plank of the hull side on each beam. To do this the keel was rabbeted on each side and mortises were cut into the rabbet at regularly spaced intervals; more were cut in matching positions in the garboard strakes. Wooden tenons were made to fit into the keel mortises, the garboard strakes then being fitted on to the tenons and pushed into a firm, close fit to the keel; holes were then drilled through the planks and tenons and wooden dowels or pins driven through to lock each joint.

The process was repeated for each succeeding hull strake, each shaped and formed to follow the curving edge of its predecessor, as well as the curvature of the hull. Jigs and spacers, as well as the eye of the experienced shipwrights, were used to guide and support the strakes as they were added. The method was extremely laborious in an age without power tools and costly in materials, since as much as three-quarters of a piece of timber could be lost in fashioning it to a precise fit.[5] The resulting hull was strong, with little built-in stress, and needed little or no caulking. Once the shell was completed approximately to the round of the bilge or the second wale, the jigs and spacers were removed and ribs, frames, stringers and thwarts were inserted to stiffen it.[6] These were secured by treenails driven into predrilled holes and copper or bronze spikes (later increasingly replaced by iron) driven into them and sometimes clenched on the inside.

Some merchant ship hulls were double-skinned and others had extra external planking as either reinforcement, patching or to support soft timber. With so much timber lost to closely spaced, sometimes even overlapping, mortises within the depth of the strake (typically between $1\frac{3}{8}$ and 4 inches (35–100 mm)[7]) a second skin could compensate for any weakness so caused. By the fourth century AD – probably earlier – the mortises became more widely spaced; no doubt it was found to be more economic in building and that the earlier extreme numbers were unnecessary as the resultant hull was just as strong. Thereafter, as the number of mortise and tenon joints continued to decline, frames and timbers became heavier to compensate and more caulking was needed. Eventually the few remaining mortise and tenon joints acted as no more than locating points for the strakes, rather than adding to strength, and although the shell was still built first, it depended upon the internal timbers for strength.[8]

Merchant ships have been found with

massive wales fitted around their hulls, secured by spikes or long bolts through the frames;[9] on warships such a wale was fitted by way of anti-ramming armour at the waterline. Two other features arose from a well-built hull using the mortise-and-tenon method; first, they proved to be long-lived, given proper maintenance.[10] To protect them from marine parasites and wood-borers, hulls were smeared or payed with pitch or tar (from natural tar ponds) or with a mixture of wax and lime.[11] During the closed sailing season, warships were pulled out of the water to be dried out, some even housed in ship sheds and annual maintenance and repair was carried out.[12] The other advantage of using the 'shell-first' method was that the hull was a strong monocoque which could absorb and distribute the shock of a ram attack. Although information about these construction methods has had to be obtained from analysis of merchant hulls, not warships, it can be assumed that the same methods were used for both. This assumption is upheld by a bronze ram recovered from the sea off Athlit, Israel to which parts of the hull that it came from, built by this method, were still attached.[13] Further evidence is provided by the remains of what are believed to be two Carthaginian warships of the First Punic War (264–241 BC) period, again built by the mortise-and-tenon, shell-first method.[14]

When they reached the northern shores of Europe, the Romans came upon a different, Celtic tradition of ship building which had evolved to suit their local sea conditions. Caesar describes the ships of the Celtic Veneti of Brittany as being heavily built, of oak with shallow keels.[15] The ships had been built without true keels but with extremely thick bottom planking to enable them to sit on a shore when the tide went out. They were built 'frame first', that is once the bottom planks were laid and secured, a skeleton of frames were set up on it and the rest of the planking nailed to them to complete the hull shell, the iron nails being bent over and clenched on the inside.[16] The Celts did not have separate warship types, but their building tradition continued in use until the end of the Western Empire and both Celtic and Mediterranean methods of building appear to have been used in parallel in northern waters.[17] It is not known whether the Celtic method was used for northern warships or only for military transports, but its inherent strength would suggest so. The problem might have been that in this form, where the strakes are not fixed to each other, they might flex and open upon a ramming impact, whereas with the Mediterranean method where each strake was locked to its neighbours, the whole hull flexed upon impact. However, whereas the latter was evolved to counter similarly built and heavier ships, the former might have proved quite sufficient not to cause undue flexing upon ramming the very lightly built barbarian craft by which they were opposed.

Materials

While taking advantage of whatever timber was locally available, shipbuilders would use different types in the manner most suited to their attributes; thus Athenian triremes had keels of oak to withstand being hauled out of the water.[18] Oak also seems to have been preferred for frames while for hulls, pine, cypress, cedar and elm,[19] together with oak, beech, larch and fir – in fact, whatever was available, even acacia.[20]

Ships were built of green, or unseasoned, wood as, especially with oak, it is more pliable and easier to work. As it contains more water however, the shipwrights had to make allowance for the change in buoyancy as it dried out with age and any possible movement as it became seasoned. In this regard, the relatively stress-free result of building with the mortise-and-tenon method was of great help, although Livy stresses the need for ships so built to be hauled out for the winter to allow drying.[21] Caesar also comments on the adverse effect upon the performance of such ships when newly built.[22] It was common practice to use green timber, indeed, during the Punic Wars, since the demand for large numbers of big ships to be built quickly left no option. Vegetius also warned against its use,[23] but recent experience has demonstrated its practicality.[24]

For masts and spars, pine and fir was pre-ferred;[25] oars were of fir.[26] The warship carried a

Although several types of sailing rig were known and used by the Romans, warships appear to have used one standard system alone. This rather stylised vessel, which could be a warship, merchant galley or even a luxury yacht, shows the full warship rig; small ships would mount only the mainmast. There is an *artemon* forward, unsupported by standing rigging, with a small, square from which a sheet can be seen. The mainmast is fully rigged with stays and shrouds. Although the latter could also represent brails; braces lead from the main yard. Above the yards, are their lifts and the spaces are filled with sailcloth colour between them and the yards and this could represent a *supparum* on each, triangular topsails. The squares on the sails represent the seams of the panels of sailcloth. Detail from a second or third century mosaic. Sousse Museum, Tunisia. (*Author's photograph*)

sailing rig of a mainmast with a large, rectangular sail, made from canvas of hemp or flax and operated rope made from hemp, papyrus or esparto grass.[27] Larger ships also mounted an *artemon*, a foremast carried well forward in the bow and raked forward, upon which a smaller square sail was rigged as a headsail. Wax-based paints were used and were available in blue, red and yellow, so any other colour could be mixed; surviving wall paintings show warships painted in a variety of colours. Paint in suitable colours was applied as camouflage,[28] and for fleet actions, on occasion, the towers of the ships were all painted the same distinguishing colour as an identification aid.

Warships carried much the same equipment as merchant ships – pumps, anchors, a skiff (*scapha*) or small boat and the like – but in triremes at least an additional device was rigged to prevent the long, slender hull from hogging or sagging. This was the *hypozomata*, a type of Spanish windlass running under the deck, connected between the stempost and the sternpost and tightened by twisting, to stiffen the hull longitudinally. There were two in the inventory of an Athenian trireme,[29] one rigged and a spare. Apart from triremes, the extent to which it was used in the larger, broader ships is not known, but its use to stiffen hulls up to the size of a quinquereme has been suggested.[30]

Mass production

With the impending Punic Wars, the Romans were confronted with the need to build large numbers of major warships relatively quickly, and thereafter to keep building, in order to replace losses; during the course of the first war, the Romans were to lose some 700 ships (of which, nearly 400 were lost to storms). To provide the numbers needed, ships were built to standard designs and specifications and mass production methods employed.[31] Livy mentions the supply of iron from Populonia, of sailcloth by Tarquinia, baulks of fir from Perugia (Perusia), Chiusi (Clusium) and Rusellae and 'timber for keels and garboards' from Volterra (Volterrae).[32] Templates and jigs would be used for the components in non-specialised workshops away from the shipyards and as Livy says, timber would be cut to set sizes and patterns at the lumber yards. The Romans built up their shipbuilding capacity in anticipation of the coming wars, probably following the appointment of the *quaestores classici* in 267 BC,

WARSHIP EQUIPMENT

Warships carried anchors as part of their equipment, Caesar refers to ships being 'anchored off'.

Above left, the lower part of an Imperial period anchor recovered from the River Rhône, showing the jointing on the wooden parts of the anchor, together with the cast lead weight which holds the assembly together. Arles Archaeological Museum. Above right, a number of cast lead anchor stocks, such as was fitted to the top end of the anchor shown on the left. The largest is approximately 4.5 feet (1.37 m) in length. Naxos Archaeological Museum, Sicily.

Left, double-action water pump. Two bronze cylinders contained pistons which slid alternately when operated by the pump handle, pumping water on both up and down strokes. Ideal for pumping out ship bilges, the type was perfected in the mid third century BC. This example, with remains of wood cladding, is later and was found at Lyons. Archaeological Museum, Lyons. (*Author's photographs*)

and having done so maintained that capacity for the duration.

The ancient sources stress the speed with which warships were built in large numbers, although apparently in unrealistic times. Thus Polybius records that at the outbreak of the First Punic War the Romans built 100 quinqueremes and twenty triremes in sixty days[33] and Livy that in 205 BC they built twenty quinqueremes and ten quadriremes from cutting timber to launching, in forty-five days.[34] If, however, the increase in shipbuilding capacity and the diversified mass production of component parts is taken into account, these reports become less far-fetched and it could be that a quinquereme

could be assembled at a shipyard in sixty days; that is, they took sixty days each to build, the same principle applying to Livy's account. Accordingly, a dedicated shipyard could build six quinqueremes per year (at two months each) and if the Romans had as few as six such yards, they could supply thirty-six ships each year. Allowing for exaggeration in the sources and that it took three months to assemble/build the ship, an average production rate of twenty-five ships per year for the course of the war would suffice to provide all of the ships known to have been deployed, even allowing for losses and wastage and as demonstrated by the analysis below.[35] Without the vast losses suffered in the

ANALYSIS OF ROMAN SHIP PRODUCTION IN THE FIRST PUNIC WAR

Year BC	Event	Built	Engaged	Lost	Plus captured	Fleet strength
267	Quaestors appointed					say 50
266						50
265		24				74
264	Start of war	24				98
263		24				122
262		24				146
261		24				170
260	Battle of Mylae	25	140	11	23	183
259		25				208
258	Battle of Sulci	25	?	?	?	233
257	Battle of Tyndaris	25	?	9	4	253
256	Battle of Ecnomus	25	230	24	44	298
255	Hermaeum/ storms	25	250	225	50	148
254		25				173
253	Lost to storm	25	30			168
252		25				193
251		25				218
250		25				243
249	Drepanum/storms	25	243	220	2	50
248		25				75
247		25				100
246		25				125
245		25				150
244		25				175
243		25				200
242		25				225
241	Battle of Aegades, war ends			12	70	283

first war, a scaled-down production rate could and did keep the fleets up to strength for the Second Punic War and thereafter.

For the imperial fleets, with the demise of the larger types, production of ships could be undertaken at the now permanent naval bases of the various fleets, each of which had shipbuilding capability. A new factor was the requirement for larger numbers of predominantly smaller vessels for the river fleets; once again, each fleet being a self-sufficient entity, had its own building facilities. Standard types and designs were used throughout the fleets; Vegetius regards the river warships as being on the 'secret list'[36] and a late authority indicates the numbers possible when in AD 412, it called for the production of 200 new ships and the refit of another twenty-five for use on the middle and lower Danube.[37]

3   SHIPBOARD WEAPONS

Rams

The earliest rams, which emerged in the mid-ninth century BC, transformed warfare on water from an encounter between armed men on vessels, to a battle between the vessels themselves for the ram made the ship that mounted it into the weapon, a guided missile no less. The earliest rams were pointed, logically enough, intended to make a hole in an enemy hull at or below the waterline and thereby cause it to become swamped and disabled. A matter to note is that although it is convenient to speak of 'sinking' ships by ram attack, insofar as our current general concept is of a holed ship plunging below the surface to the sea floor, this was not so for ancient warships. These ships were, of course, made entirely of wood, which has latent buoyancy. The heaviest item of the ship was the crew, who could be depended upon to get out of a stricken ship. That they did so is perhaps proved by exception in that there survive accounts of occasions when crews did not manage to escape, and which were notable enough to deserve special mention in the ancient sources.[1] The ram itself was a metal covering mounted upon the forefoot of the ship hull, to protect it when it came into contact with a target. The fore structure of the ship itself was designed with timbers projecting forward to carry the ram and strengthened to withstand the impact and to transmit and spread the shock of ramming back along the hull. The integrity of the hull was further protected, at least by the time of the Athenian trireme, by closing the bow of the hull behind the structure that carried the ram, forming in effect, a bulkhead.[2]

The metal covering or casting which was the ram, if it had not twisted or fallen off as a result of damage, was of insufficient weight to drag the ship down and thus a ship holed at or below the waterline by a ram attack would ship water and settle to the level dictated by the buoyancy of the wood from which it was made. It was a perk of the victor in a naval battle to tow away wrecked (or 'sunken') enemy ships, to be beached and repaired.[3] The wrecks of over a thousand merchant ships from the ancient world have been located to date, all sunk by the weight of their cargo, but not one seagoing warship has yet been found, with the exception of two Carthaginian ships found in mud off western Sicily, ironically loaded with cargoes of stone, which held them down.[4]

It was found that the early pointed rams had a tendency to become ensnared by the fibres of the edges of the timbers that they had pierced. This resulted in the attacker becoming trapped against its intended victim or in the ram being pulled off as the attacker tried to withdraw, leaving it probably as badly damaged as the victim. If it could not disengage, the attacker would in turn be assaulted by the crew of the stricken ship who had the ultimate motivation for a successful boarding counter-attack.

The ram evolved, and by the sixth century BC had become blunted at the end. Greek pottery shows it shaped in the form of a boar's head or a ram's head. The purpose of this development was to push or stove in a section of enemy hull, rather than trying to make a hole, thereby disrupting the integrity of the hull and permit the ingress of water. As ships grew in size and required larger castings for their rams, these also developed further, presumably also as a result of experience in battle. By the early fifth century BC the face of the ram had developed horizontal vanes, on either side of a vertical central spine, designed to cut through the grain and joints of an enemy's hull timbers. This form of ram, of which examples have been found,[5] was a casting with a hollow socket at the rear shaped to fit over and be fixed to the waterline wales and stempost of the ship. This was the weapon of the trireme age and obviously very

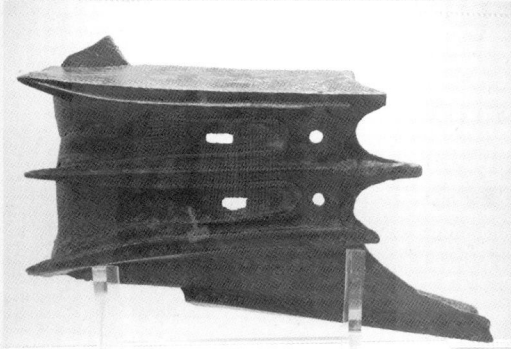

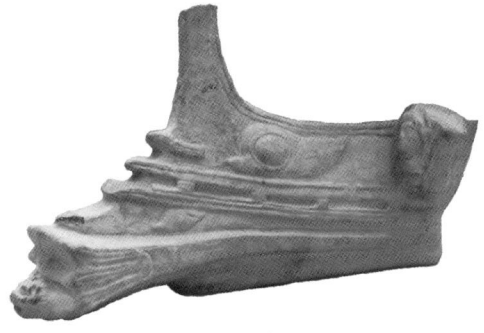

Above left, part of a bronze casting of a ram from a trireme, found off Piraeus, dated to the fifth century BC. The height of the leading edge is approximately 18 inches (360 mm). Piraeus Archaeological Museum.

Above right, stone prow of a Roman warship for use as a plinth for a statue, but showing the ram and an *oculus* or 'eye'. Second century BC. Archaeological Museum, Aquileia.

Right, detail of a terracotta rhyton or vase in the shape of a warship prow found at Vulci in Etruria, dating from the third or second century BC, perhaps depicting a Roman warship of the Punic Wars period. The lower lip of the ram, somewhat exaggerated in shape, forms the spout. The heavy waterline wales for mounting the ram and the armouring of the forepart of the ship are apparent (as is the *oculus*). Also prominent is the forward end of the oarbox, decorated with the head of a deity, heavily built to withstand attempts at a sideswipe attack. British Museum.

Part of the base structure of Augustus' victory monument at Nicopolis built to commemorate the battle of Actium. Sockets can be seen in which were mounted the rams taken from ships captured in the battle. The smallest are 40 inches across (1 m) and thought to be from quinqueremes, whereas the largest are nearly 5 feet across (1.5 m) and as high, and thought to have been taken from 'tens'.
(*Author's photographs*)

effective. The form continued in use throughout the growth of the super warships of the ensuing Hellenistic period, which in turn needed increasingly large castings to fit those ships.

As to the relative sizes of the rams fitted to warships, comparison can be made by comparing the size of their sockets; thus the ram fitted to the replica trireme *Olympias* had a socket 27.5 inches (700 mm) in width, while that of the Athlit ram, from a quadrireme, measured 33 inches (840 mm). Many rams from different sizes were taken from captured ships after the battle of Actium and mounted on Augustus' victory monument at Nicopolis; the sockets there range in size from 40 inches (1,020 mm) to a massive 5 feet (1.51 m) in width, some of the mounting sockets being over 5 feet 6 inches (1.7 m) in height. The smallest of these is considered to be from a quinquereme, with the largest from a *dekares* or 'ten'.[6]

With the elimination of other navies, the Imperial navy's opponents changed from armoured hulls to the lightly constructed open ships of the barbarian raiders and the shape of their rams changed to deal with this different type of ship. By the early second century AD an upward curving, scythe-like ram had been developed, sheathed in metal but without the blunt, vaned end[7] and intended to ride up and over an enemy bulwark and crush or submerge it, rather like an icebreaker breaking through ice. This form remained in service up to the end of the western empire and beyond. There is some evidence to suggest that a modified form was used by the northern fleets, being a short, possibly square in section, pointed ram. In the rougher northern waters, such a form could serve either to ride over and swamp or to pierce an enemy, doing damage at whatever point it struck.[8] Even in the Mediterranean however, the sea conditions and/or violent battle manoeuvres could cause a ram attack to hit the side of an enemy anywhere from below the waterline, if heeled away, to as high as the thranite oarsmen.[9]

Fire weapons

Fire weapons were, unsurprisingly, little used by ancient warships, which were made of wood

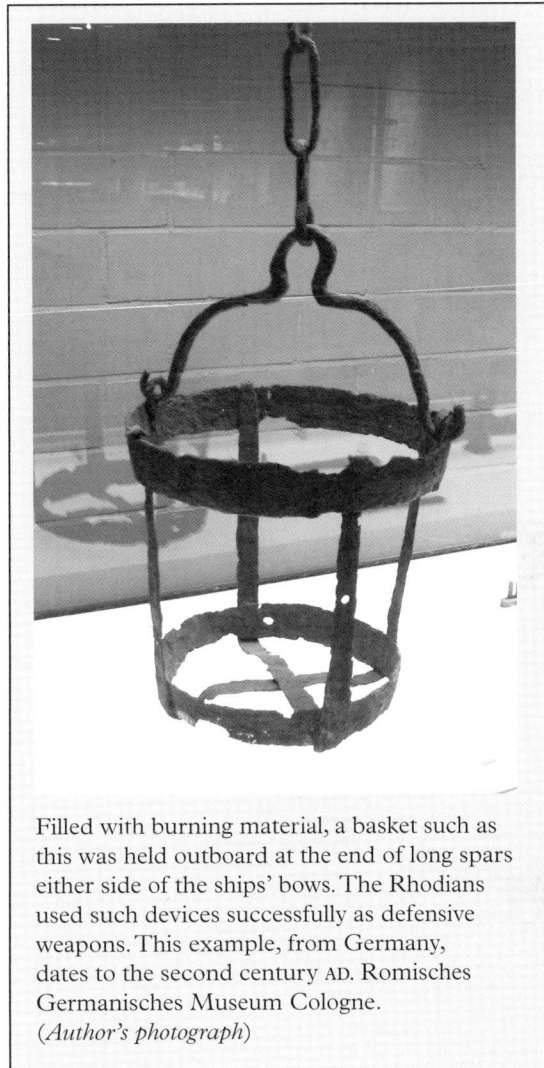

Filled with burning material, a basket such as this was held outboard at the end of long spars either side of the ships' bows. The Rhodians used such devices successfully as defensive weapons. This example, from Germany, dates to the second century AD. Romisches Germanisches Museum Cologne.
(*Author's photograph*)

with natural fibre cordage, their hulls payed with wax and pitch,[10] in a Mediterranean summer they could hardly be more flammable. There is no record or evidence of cooking facilities aboard the ships, although small oil lamps were carried for navigation and signalling. In 190 BC a Rhodian squadron was trapped in a harbour near Ephesus by the Syrian fleet of Antiochus. The latter landed a strong force of troops behind the harbour, forcing the Rhodians to man their ships and attempt a break-out from the harbour. The superior enemy fleet was waiting to pick them off piecemeal as they emerged from the narrow harbour mouth and although twenty ships were lost or captured, seven or more hung flaming braziers from long booms on either side of their bows, ready to drop them on to any enemy ship that ventured too close; they dared not and those ships so equipped escaped.[11] It seems that the Romans and Rhodians had experimented with the braziers during the preceding winter but decided that the risks outweighed the benefit and their use by the Rhodians in this instance was driven by desperation. There is a first-century BC graffito at Alexandria of a ship with such a brazier rigged outboard on a boom, but this may well be a navigation light or improvised beacon, held well away from the ship.

As to the use of flaming projectiles, although fire arrows had long been used on land, the first indication of their use at sea was in 42 BC during the civil wars. A troop convoy, escorted by some triremes, had been caught crossing the Adriatic, on its way to join Octavian and Antonius. The wind failed and the transports tied themselves together to form a stable fighting platform for the troops. Their opponents simply stood off and shot fire arrows, burning many of the ships, the remainder cutting themselves free and surrendering.[12] The next occasion (that we know of) when fire arrows were used was at Actium. Seeking a conclusion of the, thus far, evenly matched battle, Octavian sent ashore for burning material; this his men used in fire arrows or tied to javelins and even brought their ships close enough to throw lighted torches at their enemy. At longer range, they used artillery to hurl pots of blazing charcoal and pitch. The fires caused on Antonius' ships were contained until the wind strengthened and many of his ships were consumed by fire.[13] It remains notable that there were no fire weapons aboard the ships and that Octavian had to specifically order them brought from his encampment ashore. By the late Empire, Vegetius[14] warns of the danger of 'arrows wrapped in burning oil, tow, sulphur and bitumen' and shot from catapults; however, there are no accounts of the use of such weapons in the reported battles other than at Actium, nor does Vegetius say that theses missiles were carried aboard ships, rather than shot from the shore.

The other way that fire was used as a weapon afloat was by the use of fireships, recorded as used by the Syracusans against the besieging Athenian fleet in 413 BC: 'seeking to burn the remainder of the fleet, they loaded an old merchant vessel with faggots and brands, lighted them and let the ship go, the wind was blowing right on the Athenians.'[15] This defines the use of a fireship, although in this instance, the intended victims managed to push it clear.

The Romans made use of this tactic in their civil wars and in 48 BC, an admiral of Pompeius', having brought with his fleet ships filled with pitch, pine resin and other flammable materials and having a favourable wind, launched them against a Caesarian fleet laying in Messina. The attack succeeded, all thirty-five ships of Caesar's fleet being burned to the waterline. A second attack by forty fireships against another Caesarian squadron, more alert than the former, burned five ships, but Pompeius' fleet was itself routed by the counter-attack.[16] Finally, in the last operation by a joint western and eastern fleet against Vandal-held Carthage in AD 457, their closely packed fleet and transports were held against the shore by an onshore wind. Taking advantage of those ideal conditions, the Vandal fleet launched a fireship attack, following it with a ram attack and destroyed half of the Roman fleet.[17]

Towers

The earliest surviving account of the mounting of a tower on a ship was by the Athenians, who built one on a merchant hull to assist in their siege of Syracuse in 413 BC.[18] Alexander the Great did the same for his siege of Tyre in 332 BC, as did Marcellus for his assault on Syracuse in 214 BC.[19] In all of these cases, the ships were taken, rowed or towed, to their desired station and not intended to otherwise move. At the beginning of the First Punic War, the Romans devised a lighter form of tower with a timber frame, covered by canvas and which they mounted on the decks of their larger warships. Quadriremes were the smallest types to be so equipped; towers could be mounted forward, aft or amidships and even in pairs, one fore and one aft, on the biggest ships.[20] It may be that towers were also mounted on the very largest of ships, athwartships in pairs, jettied out beyond the hull on each side. The evidence for this, in the absence of extant illustrations, is in Polybius' account of the battle of Chios in 201 BC between the fleets of Philip V of Macedon and Rome's allies, Pergamum and Rhodes. At one stage in this great battle, the ship of the Pergamene admiral Dionysodorus, a quinquereme, attacked a Macedonian 'seven', missed and brushed along the side of his intended victim. The latter had quickly withdrawn their oars but Dionysodorus' ship was not quick enough and had its starboard oars sheared off and also 'had the timbers supporting his towers smashed to pieces . . .' This was of course impossible for towers mounted amidships.[21] Up to six archers, slingers or javelinmen could be accommodated on each tower, able to shoot down on to an enemy deck.

The towers, as a permanent feature, were heavy enough and added to top weight, wind drag and thus instability to the ship, and undoubtedly contributed to the loss of so many ships in storms. They nevertheless continued in use until the thirties BC, when a new type was introduced by Agrippa. This was of lighter construction, collapsible and able to be stowed flat on deck when not in use, to reduce top weight. The canvas coverings were often painted a distinguishing colour on all the ships of a squadron to aid in identification; by the later civil war period, they were sometimes painted to resemble masonry.[22]

Boarding

Before the Punic Wars Roman naval activity had for the most part been carried out by a modest numbers of smaller ships, against like opponents and boarding an enemy vessel had been an ad hoc affair. In the great wars, large numbers of big ships carrying tens of thousands of men were engaged and pitted against an enemy, initially superior in seamanship to the majority of the Roman ships. Additionally, far larger contingents of marines were carried

who, to be effective, had to have some means of crossing to an enemy deck, while themselves being under attack.

If ships lay close alongside each other, marines were able to clamber across the gap along the ship's length.[23] The Carthaginians preferred ramming attacks and so the gap across their bows was too great to jump. The Roman solution was to mount a 'bridge' 36 feet long (11 m) and 4 feet wide (1.2 m) with a slot in one end which fitted around a post 24 feet in height (7.3 m) mounted in the foredeck; the bridge was hauled up by rope and pulley, to be held almost vertically against the post. At the other, free end of the bridge was a metal spike (the actual *corvus*, raven's beak) the slot allowed the bridge to be adjusted as to length and swung from side to side, so that when the Roman ship came within range, the bridge was dropped across the enemy deck, into which the *corvus* embedded itself, locking the ships together.[24] The bridge was wide enough for men to cross two abreast and, covered by missiles shot from their towers, the Roman marines could swarm across. The bridge had a low rail on each side to knee-height and the soldier's shields covered them from knee to shoulder, so is was preferable to pass starboard to starboard of the enemy and

A lower tower mounted forward on a Roman warship. Detail from a wall painting, first century BC, showing a sea battle. Palazzo Massimo, Rome. (*Author's photograph; see also plates III and IV*)

Iron spring-mounting frame, from a torsion-spring artillery piece, with an iron spring washer and retaining bar. The sinew spring was threaded through the washer and over the bar, the washers at each end being held at each end of the frame. A recess on one side allows extra forward travel for the bow stave and four brackets are on the sides of the frame, enabling it to be secured to the field frame of the machine. Possibly from a 'three-span' (i.e. shooting a 27-inch (640 mm) bolt) called a *cheiroballista*. Early second century AD. The iron bolt-head is of a type used in the piece. Lyons Archaeological Museum. (*Author's photographs*)

drop the *corvus* so the marines would cross with their shielded sides to the enemy. The equipment was bulky and could only be mounted on the big quinqueremes.

Although outstandingly successful in action, the drawback was that the *corvus* was heavy and mounted high in the very bows, it added considerable top weight and adversely affected sea-keeping qualities. The loss of so many ships equipped with it in storms led to the device being discarded by about 250 BC and being replaced by an alternative form of 'boarding bridge'. No description of this device survives but it was lighter and able to be dismantled and stowed on deck, or even jettisoned if necessary, in effect a lightweight *corvus*, able to be manhandled without the need for the heavy post and tackle.[25]

Artillery

Artillery was a feature of ancient warships almost from its invention in the early fourth century BC. At that time the engineers of Dionysius, the tyrant of Syracuse, made a large composite bow, larger and more powerful than could be drawn by a man, this they mounted on a timber stock, together with a winch and trigger mechanism to draw and release it. The whole was connected to a stand by a universal joint which enabled the machine to be trained both vertically and horizontally and thereby aimed. The stock to which the bow was fixed had a central channel along the top in which was fitted a slider with the trigger mechanism and a channel for the missile; this was locked to the bowstring and the whole slider pulled back by the winch, checked by a ratchet along the sides of the stock. Upon release of the missile, the slider was pushed forward, to re-engage the bow string and winched back again to reload. These machines could hurl a large arrow or, with an adaptor, a stone shot, up to 300 yards (275 m). Mounted on ships, they could cause devastation if shot among an enemy's rowers, easily piercing the leather screens which had been sufficient to stop javelins or arrows and which had provided their protection up to then; the advent of artillery led to the progressive boxing in of the rowers, to provide them with sufficient armour protection.

Reliable and simple to maintain, these machines were in use until about 240 BC; before that, the engineers of Philip II of Macedon (reigned 359–336 BC) had perfected a new type of propulsion system for artillery. The composite bow was replaced by two short bow staves, each inserted into a skein of animal sinew which had been woven into cords, stretched, oiled and twisted to form torsion

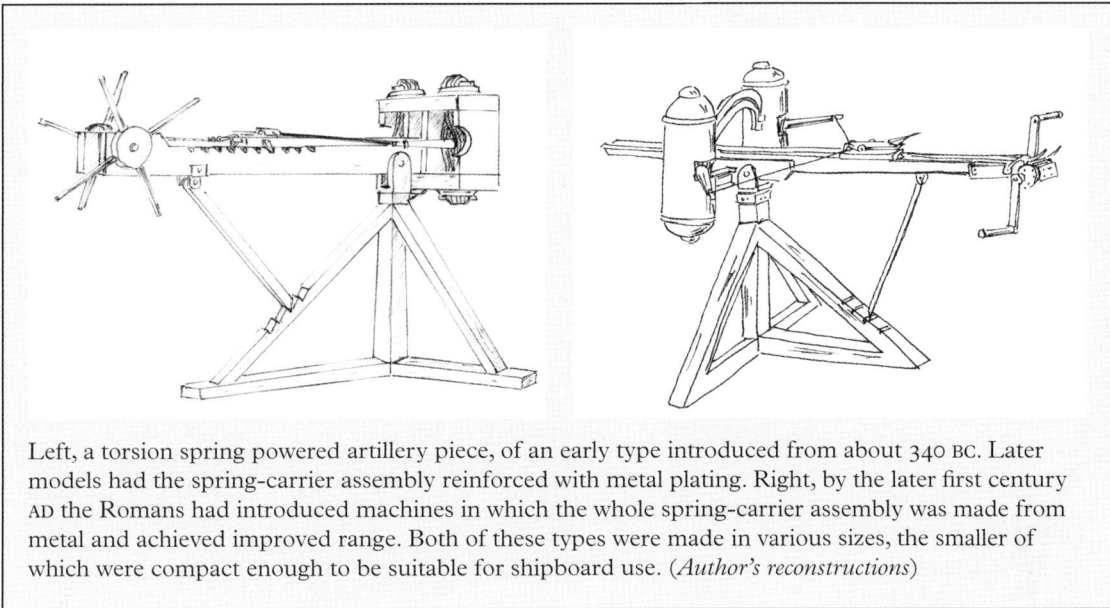

Left, a torsion spring powered artillery piece, of an early type introduced from about 340 BC. Later models had the spring-carrier assembly reinforced with metal plating. Right, by the later first century AD the Romans had introduced machines in which the whole spring-carrier assembly was made from metal and achieved improved range. Both of these types were made in various sizes, the smaller of which were compact enough to be suitable for shipboard use. (*Author's reconstructions*)

springs which stored great power. These were mounted in special carriers and kept in tension by passing them through a hole and washer at the top and bottom and retained by a bar; they were mounted each side of a stock with slider and trigger, similar to their predecessors. Being a good deal more powerful, range was increased to some 400 yards (366 m) and they could be made in ever increasing sizes. The springs were susceptible to damp and needed to be regularly removed and restretched and oiled, requiring in turn, specialised artillery artificers. As the pieces were made by estimating, with no set pattern of parts, performance could vary greatly, which is why the earlier form continued in use.

From about 275 BC Ptolemy of Egypt's engineers had developed a formula for building torsion spring artillery pieces. Starting with the length of arrow that the intended machine was to shoot, the formula dictated the fraction of that length (it was 1/9th) which was to be the diameter of the holes in the carrier through which the springs would pass. All dimensions of all of the components of the machine were then dictated by the formula, as multiples or fractions of that hole diameter. These 'formula machines' proved to be reliable, of known performance and could be mass produced in standard sizes.[26]

These machines supplanted the earlier types and became standard equipment aboard warships. The smallest type seen to mount artillery was the trireme, whereas quinqueremes and larger types could carry up to ten. The larger ships could also carry larger sizes of catapult. These weapons, although capable of inflicting damage to crew and ship, were not 'ship destroyers',[27] and the modestly sized and well attested three-span (shooting an arrow 27 inches in length (685 mm) with a stock length of 4 feet 6 inches (1.27 m) and two-cubit (3-foot or 915 mm arrow) with a stock length of 6 feet (1.83 m) machines would have been ideal as anti-personnel weapons.

Apart from shooting at each other,[28] warships could bring their artillery to bear in support of an opposed troop landing, as Caesar's ships did in Britain in 55 BC[29] and again in his attack on the Heptastadion at Alexandria in 47 BC.[30] Artillery on ships could also be used as floating batteries, as was done to cover the building of a bridge over the Euphrates in AD 62.[31]

Apart from the normal complement of arrow and stone shooters, larger machines were on occasion, mounted for siege or assault work, especially during the Punic Wars. Siege artillery was substantially larger and predominantly stone throwers, the very biggest being capable of hurling a stone shot weighing an amazing 260 pounds (118 kg). Although siege machines smaller than this were mounted, the additional weight of the machine and of its stone shot had to be considered. The weight of a three-span piece has been estimated at about a hundredweight (112 lb, 61.7 kg) and that a quinquereme could carry ten, plus a couple of small stone throwers and still have its normal complement of forty marines.[32] Clearly the installation of such machines had to be offset by weight savings, the marines being replaced by artillerymen and the rowing crew being reduced to just enough to manoeuvre the ship into position and all stores landed.

As examples of this practice, in 210 BC, the Roman fleet supported an assault upon Naupactus from seaward using siege artillery.[33] A year later, in 209 BC, both war- and merchant ships were so equipped for the siege of Taranto[34] and warships with extra artillery joined the assault on Cartagena.[35] Finally, Scipio put siege artillery aboard his warships for a feint against Utica in 204 BC.[36]

Grapnels attached to a line and thrown, had long been in use to ensnare and pull an enemy ship in so that it could be boarded[37] but for the war against Sextus (38–36 BC) Octavian's admiral Agrippa introduced the *harpax* (harpoon), a grapnel attached to the end of a shaft 7.5 feet in length (2.3 m.) and lined with metal strips so that it could not easily be cut through.[38] It was shot from the larger onboard catapults and trailed a line which could be hauled in. Although the range of the machine was considerably reduced by the added weight of the missile and the drag of the line, it could still far outrange any hand-thrown grapnel.[39]

Artillery was not mounted on towers which, being comparatively flimsy, could not support the weight or the shock of discharge of the machines. Further, the position of the universal joint mounting, close behind the spring-

Grapnels such as this, attached to lengths of rope, were thrown to ensnare an enemy ship and to pull it close enough to enable it to be boarded. A variation of this was attached to a shaft (the *harpax*) which could be shot from an artillery piece for longer range. Iron grapnel, second century AD. Romisches Germanisches Museum, Cologne. (*Author's photograph*)

carrier frame, prevented the piece from being depressed to shoot down on to an enemy deck. Low platforms for artillery were sometimes erected on deck to give an elevated position and enable them to shoot on either beam.[40]

In the late first century AD there was a radical redesign of the smaller types of artillery pieces which resulted in replacing the former wood and iron spring assemblies with finely made all-metal parts for the torsion spring carriers and their mounting frame. Although more complex and requiring more specialised manufacture, they could be dismantled and worn or damaged parts quickly replaced, the parts being standardised and interchangeable. They were more compact and also more powerful, range being increased to in excess of 500 yards (457 m). These are the machines so much in evidence on Trajan's Column. Light, compact and powerful, these machines were ideal for mounting on warships, including many of those too small to have carried the earlier types.[41]

4 SHIP TYPES – THE REPUBLIC

Rowing systems

The prime motive power of the ancient warship was provided by oars; sailing rigs were carried and used under favourable conditions to augment the rowers; wherever possible, the rig was put ashore before a battle. The earliest warships were little more than elongated rowing boats with one man per oar, all sitting at the same level (monoreme) one behind the other on each side, plying their oars across the bulwark. To increase power and speed, additional pairs of rowers were added until at twenty-five pairs (fifty oars) the hull became too long and narrow, making it structurally weak and unmanoeuvrable.

The only way to increase motive power is to increase the number of oars per given length of hull and so, in the eighth century BC[1] a second set of rowers was added to an enlarged hull, sat a little above and outboard of the original group and utilising the spaces between them. This doubled the motive power in the same length of hull or, preferably, enabled the same number of rowers to be accommodated in a shorter, more handy hull (the bireme). It was, of course, possible to build a beamier hull and have each, longer oar worked by two men seated on the same rowing bench or 'bank', i.e. double-banked. This would not, however, double the power as the oar stroke is dictated by the reach of the inboard man, the outboard man only being able to pull through a part of that arc; in the bireme, all rowers could contribute their full stroke.

The Corinthians are credited with the invention of the trireme in the late sixth century BC.[2] A third level of rowers was superimposed above and outboard of the other two, rowing across an outrigger mounted along the side of the hull. The total number of rowers and therefore propulsive power, was increased to 170, more than doubling the previous maximum, but in a hull that was not greatly larger. It was a leap

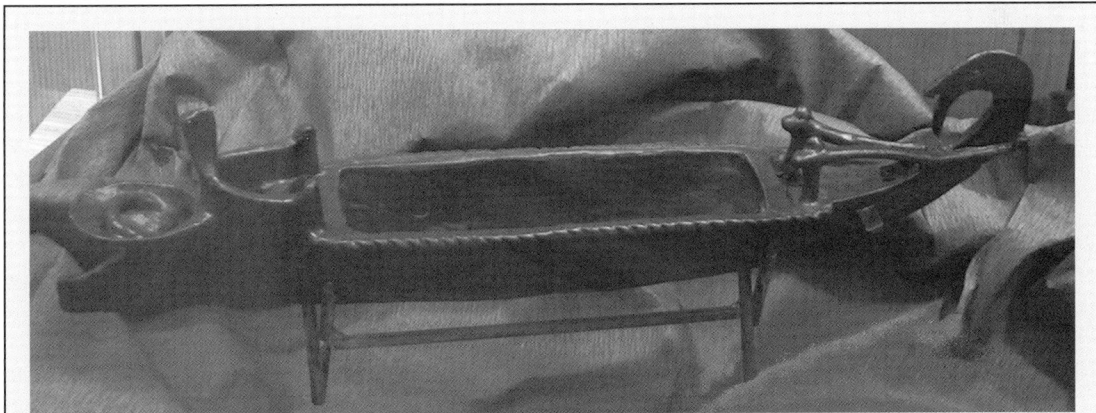

What did the first Roman warship, recorded in 394 BC, look like? For their earliest naval operations the Romans had only a few, small warships; although slightly earlier, this fifth century BC representation of a monoreme penteconter illustrates that kind of vessel. This is a copy of a bronze oil lamp in the shape of a warship, hence the bowl above the ram, but otherwise is fairly accurate. It shows the raised fighting platform in the bows and the steering and command deck at the stern, with a bracket to support the port-side rudder. (*Author's photograph*)

A copy of the 'Lenormant Relief'. A fragment, found in Athens, of an ancient relief of about 400 BC, showing part of the starboard side of a trireme accurately and to scale. It clearly shows the three levels at which oars are being operated, as well as the stanchions that support the outrigger and upper deck, together with the thranite rowers of the topmost level. It has proved an important piece of evidence in enabling the reconstruction of the ancient Athenian trireme. (*Author's collection*)

in warship development, being considerably faster and more powerful than and able easily to overawe its predecessors. It was the standard warship of the Persian and Peloponnesian wars of the fifth century BC and it remained in service with various navies and in changing forms up until its last recorded appearance at the battle of the Hellespont in AD 323.

The earlier warships were known as conters (*kontoroi*, singular *kontoros*, Greek) and they were classified by prefixes identifying the number of oars, so a penteconter (*pentekontoros*) was a ship of fifty oars.[3] The word reme (*remus* is the Latin word for oar) has been adopted as a convenient term for a number of rowers all plying their oars at the same, horizontal level in a ship, thus monoreme, bireme and trireme. The ancient authors did not, however, distinguish between monoremes and biremes, commonly referring to them all as penteconters. One definite point is that none of the authors, nor any iconography, refer to or show more than three remes. Of these the topmost are known, in their Anglicised version as thranites, the middle as zygites and the lowest level as thalamites.

Ships later grew beyond the classical trireme, with the development of the 'five' (quinquereme) and 'four' (quadrireme) and so on in a range from three to forty, the largest type known to have been used in battle being a 'sixteen'. The number refers to the longitudinal files (*ordines*)

of rowers on each side of the ship, as well as to the number of men operating each vertical group of oars. With no more than three remes and the maximum practical number of men able to operate a single oar limited to eight,[4] multiple manning of each in some fashion is implied and any combination of three oars and up to eight men per oar is possible. The largest types were developed by the Hellenistic powers of the eastern Mediterranean but were not adopted by the Romans and Carthaginians, who, apart from a few 'sixes', standardised their fleets on the quadrireme and quinquereme

Early types

There were three early influences in Roman warship design, the Etruscans, the Carthaginians and the Greeks of Italy. The Etruscans, 'strong on land and at sea very strong indeed'[5] had dominated Rome; the fledgling Republic had entered into treaties with Carthage and as it expanded, the city had absorbed the Greek settlements. These close contacts kept the Romans well aware of and familiar with the types of warship in use and of naval developments and their earliest warships reflected those influences.

Nevertheless, having only a limited coastline and thus needing only a few ships for coastal patrol, the earliest Roman warship requirement would have been met by the smaller conter types.

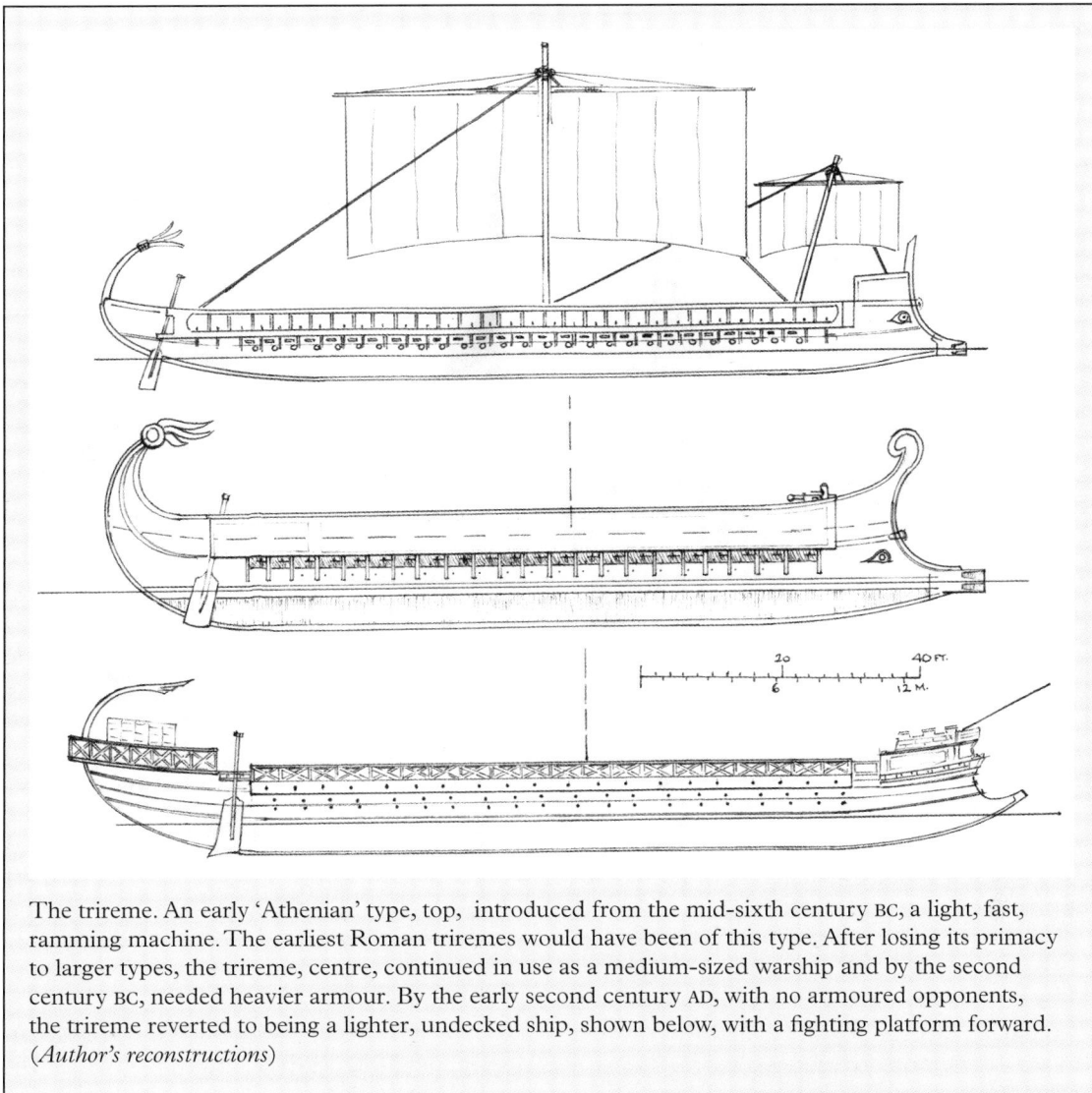

The trireme. An early 'Athenian' type, top, introduced from the mid-sixth century BC, a light, fast, ramming machine. The earliest Roman triremes would have been of this type. After losing its primacy to larger types, the trireme, centre, continued in use as a medium-sized warship and by the second century BC, needed heavier armour. By the early second century AD, with no armoured opponents, the trireme reverted to being a lighter, undecked ship, shown below, with a fighting platform forward. (*Author's reconstructions*)

The conter was ship with a ram and propelled by between twenty and fifty rowers, one man per oar and seated along each side of the ship. The ship could be arranged as a monoreme, with all of the rowers at the same horizontal level, one behind the other along each beam; or as a bireme, with some of the rowers sitting at a higher level, slightly above and outboard of and utilising the spaces between the lower level of rowers, doubling the motive power of the same size of ship or enabling the same number of rowers to man a shorter, more handy hull. The ships would usually have a deck forward for the fighting men and another aft for the steering and ship handling; in between could be open or with a partial deck to bridge the bow and stern, or be completely decked over to protect the rowers. They were able to mount a single mast just ahead of amidships, carrying a yard with a single, rectangular sail and were steered by twin large oars or side rudders mounted aft on each side. The crew required for such ships was, of course, one man per oar, together with a captain and his lieutenant, a few sailors to handle the sailing rig and anchors and a number of marines and archers as the fighting component. To what

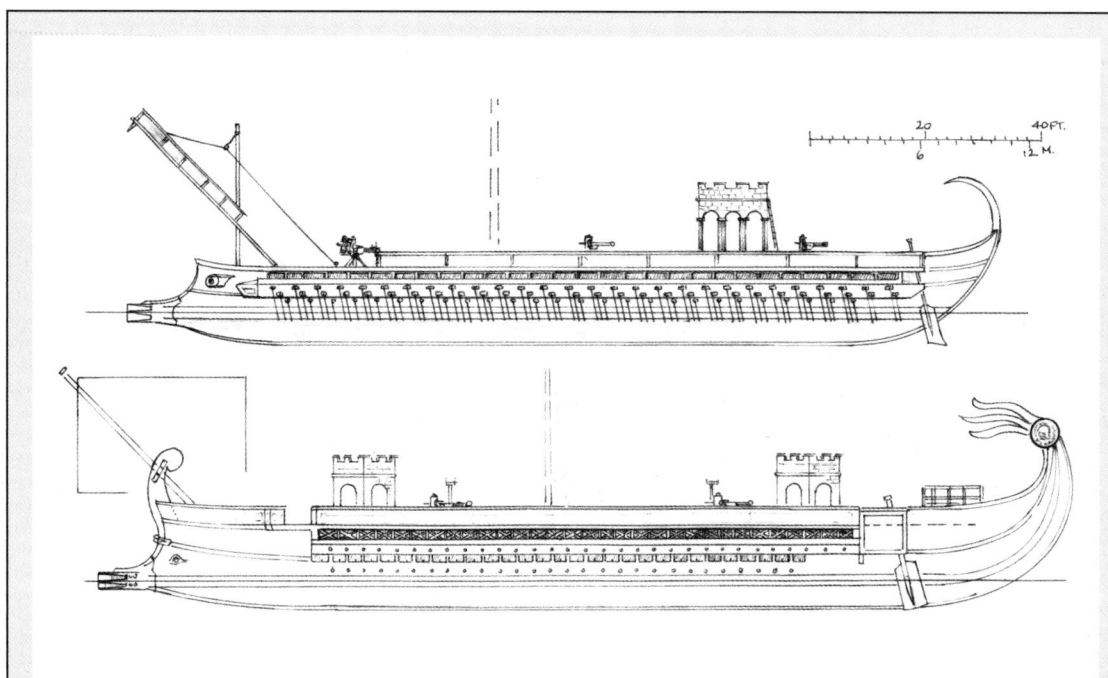

The quinquereme was the most prominent warship type of the Punic Wars and of Rome's subsequent naval operations in the Mediterranean. Top, the early type that gave the Romans their early victories over the Carthaginians, but which suffered heavy losses in storms. The ship is short, blunt and mounts a heavy tower aft, together with the *corvus* and its associated tackle forward. Below, the type introduced later in the First Punic War, with finer lines, lower towers that could be dropped to deck level and a lightweight boarding bridge also stowable on deck. (*Author's reconstructions*)

degree arms were carried for the rowers in the event that they had the opportunity to join in a fight is not known but, where such ships were cruising with piratical intent at least, every men aboard was combatant. The suggested total crew required for a twenty-oared ship might therefore be two officers, twenty rowers, four sailors and half a dozen marines; for a fifty-oared ship, three or four officers and leading rates, fifty rowers, six sailors and perhaps ten marines: totals of thirty-two and seventy men respectively.

With the continued expansion of Roman power, including their acquisition of an Adriatic coastline in 304 BC, their requirement for more ships with which to protect and control their coasts grew and by this time, although there is no direct evidence of the composition of the fleet, it can be reasonable assumed that they had started to operate triremes. The primacy of the trireme in the battle line had been overtaken in the early fourth century BC by the evolution of larger, more powerful types. The trireme remained the fastest of the ancient warships and continued in widespread use for scouting and any duty where speed was desirable; its continued value is emphasised by the order to build twenty prior to the outbreak of the First Punic War, as part of the Roman fleet expansion programme and to add to those already in service.[6] The Punic Wars provided the incentive for Rome to build a battle fleet of the latest and larger types of ship and saw the introduction of the quadrireme, the quinquereme and *sexteres*, as well as new, smaller types, although earlier types continued in service in subsidiary roles.

Battle fleets

The most widely used type, which formed the mainstay of the fleets of both sides, was the quinquereme or 'five', a type perfected at Syracuse in Sicily and first reported there in 398 BC.[7] The Carthaginians adopted the type and, to oppose them, the Romans ordered 100 to be built.[8] It may well be that the Romans had a few of these ships prior to the wars but those now built for it were heavily built and somewhat cumbersome compared to their Punic equivalents; later ships were built to a lighter, faster and more seaworthy design.[9]

As a type, the quinquereme was fully decked ship some 150 feet (47 m) to 170 feet (51.5 m) in length overall, with a beam of between 22 feet (6.7 m) and 26 feet (8 m) and a deck some ten Roman feet above the waterline (9 feet 3 inches/ 2.96 m).[10] There were up to 300 rowers[11] and forty marines, as well as sailors and officers, in all, nearly 400 men. The ship was big enough to mount the *corvus,* towers and artillery. The most likely arrangement of oars was as a trireme, with two men per oar for each of the upper two remes and a single-man oar at the lowest (thalamite) level. Alternatively, the ship could have been a bireme with triple-banked oars above double- banked lower oars; this, however, would result in a deck at the same height above water as that of a quadrireme whereas the deck of a quinquereme was known to have been higher.[12]

Although the quinquereme was the work-horse of the fleet, the Romans also built and employed a small number of quadriremes, in addition to captured Punic ships which they recommissioned. This type was evolved by the Carthaginians[13] and was most likely to have been an enlarged bireme, with each oar double-banked. The ships were about 140 feet (42 m) in length, with two remes of between twenty and thirty oars on each beam.[14] Although generally inferior in the battleline to the quinquereme, the quadrireme was handy, a good sea boat, quite fast and economical to operate; it was also the smallest type to be able to mount towers and some artillery. It was used in battles by both sides and was favoured by the Rhodians, who used it to great effect.

The largest type adopted into regular Roman service was the *sexteres* or 'six'. It is not known if these very big ships were built by the Romans (there is no reason why they should not have been) or captured from the Carthaginians. Being comparatively cumbersome and expensive in manpower, their numbers remained very small, for example, two are recorded as the flagships of the Roman fleet at the battle of Ecnomus in 256 BC.[15]

The rowing system of these ships is open to conjecture; Polybius mentions a Carthaginian 'seven' as a monoreme, rowed by seven men to each oar[16] but this may well be a later misinterpretation to accord with the more familiar Renaissance *a scaloccio* system which could have seven men on an oar. Given that these ships were very large, with towers, artillery and perhaps as many as 150 marines and very high decks, floating fortresses in fact, to utilise the power of their rowers, they were of bireme or trireme configuration. As a bireme, with three men per oar or as a trireme with two men to each oar, the ships would employ between 250 and 300 oarsmen.

At the other end of the scale, the period also saw the introduction of the *liburna.* Operations along the Dalmatian coast led to the Romans capturing large numbers of the local native craft, called *lemboi* by the Greeks. Appian describes these craft as 'a lightly built open galley without a ram, carrying about fifty men and rowed in a single reme'[17] and 'of not more than sixteen oars' (presumably per side).[18] Found to be useful as a light warship, the boats underwent considerable development by the Romans, who went on to produce a bireme version with a ram.[19] These liburnians quickly replaced the older types of light warship, going into widespread service and in time becoming the best-known Roman warship type. A typical ship would be about 80 feet (24 m.) long, rowed as a bireme by between fifty and sixty oars, with one man to each oar.

Pax Romana

After the battle of Actium, with the surrender of Antonius's remaining ships and the defection of Cleopatra's Egyptian fleet, the Roman navy was left without any naval force that could oppose it. Apart from the few ships of client kings and states, who could form no serious naval threat, the only warships were Roman and all under the command of Octavian, now the undisputed ruler of the Roman world. Without any possible adversary, the great fleets of quinqueremes became pointless and the very different duties that would be required of the newly forming imperial fleets could be satisfied by smaller, more economical ships. A few quinqueremes and at least one *sexteres* were kept in service, mainly for ceremonial duties and to keep the technology used in building them alive, they were not replaced when decommissioned and the types disappear from history by the late first century AD.

The workhorses of the imperial fleets were the quadriremes, triremes and liburnians,[1] the first two now effectively the largest types in fleet service in substantial numbers. The quadrireme, continued in service in some numbers at least into the mid-third century AD. It disappears from the record, at least in that recognisable form, before the end of that chaotic century. The trireme had evolved from its fifth-century BC heyday, becoming more heavily armoured, decked overall and able to mount some artillery by the late first century BC. Without opposition, it could revert to its lighter, faster origins and it further evolved so that by the early second century AD, it appears on Trajan's Column as an open, undecked ship. Against enemies without artillery, light screens to protect the rowing crew were sufficient, augmented by a prominent fighting platform in the bows. The trireme continued in widespread use until AD 323, when the fleet of the Eastern emperor Licinius'

with as many as 200 triremes, was soundly defeated by the fleet of Constantine composed of a new type, called by the then archaic name of '*triaconter*'.[2] The classical trireme thereafter disappeared from the historical record.[3]

The word liburnian, although probably the best-known term for a Roman warship, is an inexact term which seems to include a number of variations of the basic light warship, there is no positive identification and the classification probably included several variants and was in use for light warships generally. The light warship appears from depictions to have been an open bireme of fifty to sixty oars, with a ram. By the early second century AD, if not before, the ship appears, again from Trajan's Column, to have evolved into river and seagoing versions.[4] In its various forms, the liburnian continued to serve into the mid-third century and beyond.

To supplement the liburnians, scout ships were used, designated as *speculatoriae* or *scaphae exploratoriae*. Although the classifications are clear, what is not is whether they referred to a distinct class of vessels or were applied to any ship employed on scouting missions. Light craft are commonly shown by the iconography, although none can be definitively identified as 'scouts'. Nevertheless, upon the assumption that a distinct type was in use, a small open, monoreme ship of twenty to thirty oars, with a light sailing rig would accord with the evidence, as well as with the requirement for a small scouting vessel.

The river frontiers of the Empire, the Rhine and Danube, engendered the development of types of ships and boats optimised for river operations. In addition to the river liburnians, the columns of both Trajan and Marcus Aurelius in Rome show many representations of a standardised military river transport in use on the Danube. These ships were rowed by oars across the bulwark, or even towed and are estimated

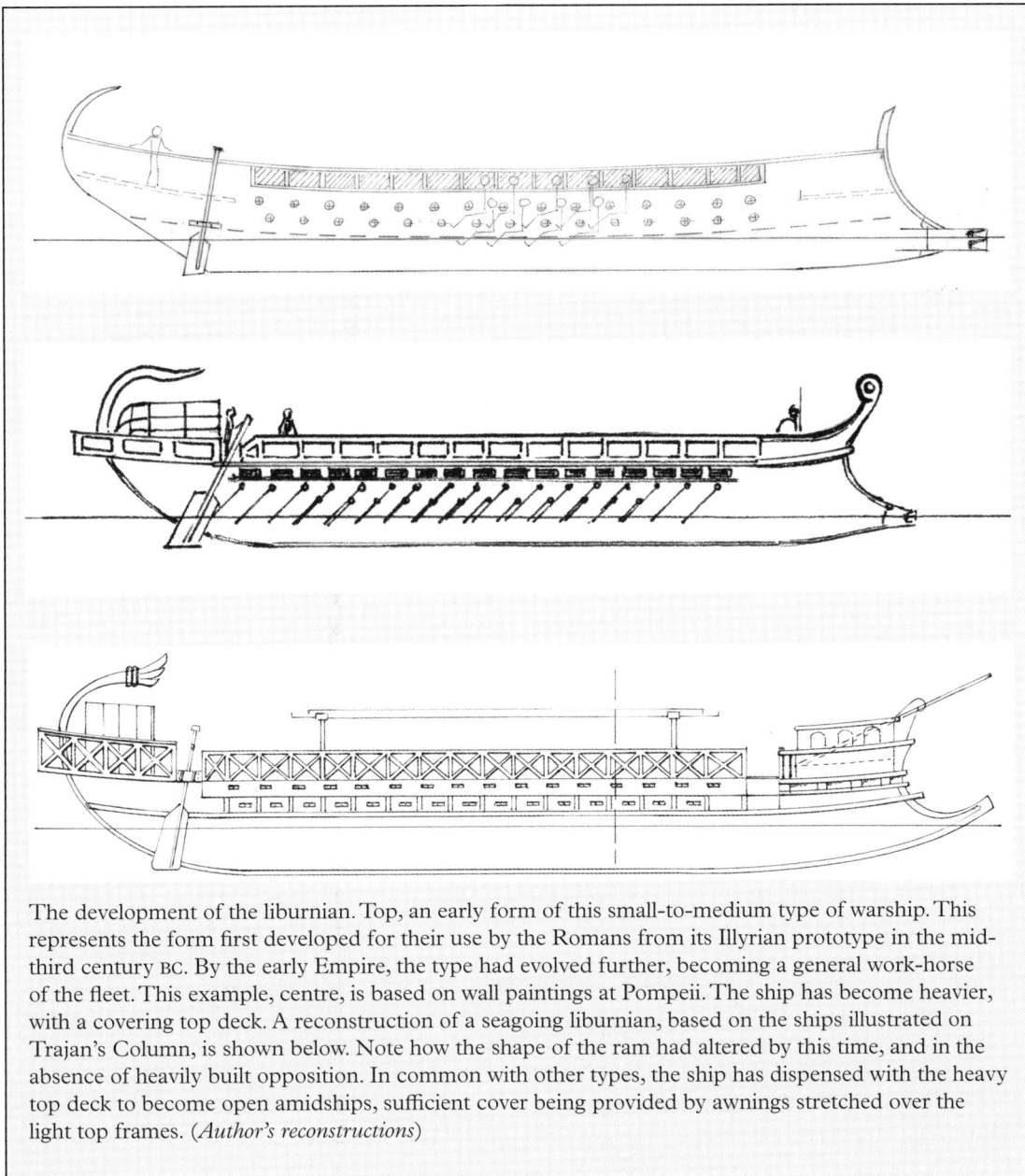

The development of the liburnian. Top, an early form of this small-to-medium type of warship. This represents the form first developed for their use by the Romans from its Illyrian prototype in the mid-third century BC. By the early Empire, the type had evolved further, becoming a general work-horse of the fleet. This example, centre, is based on wall paintings at Pompeii. The ship has become heavier, with a covering top deck. A reconstruction of a seagoing liburnian, based on the ships illustrated on Trajan's Column, is shown below. Note how the shape of the ram had altered by this time, and in the absence of heavily built opposition. In common with other types, the ship has dispensed with the heavy top deck to become open amidships, sufficient cover being provided by awnings stretched over the light top frames. (*Author's reconstructions*)

to have been approximately 48 feet (14.6 m) in length and 16 feet (4.9 m) beam;[5] they also had the facility to be used, rafted together, as pontoons for bridging. The two columns, the former built after AD 106, the latter shortly after AD 180, indicate that the type was in service for nearly a century, in substantial numbers.

To complement the river liburnians, lighter warships were developed for patrol and interdiction on the rivers from at least the late first century AD. From examples recovered[6] they were open, monoreme craft of up to 50 feet (15.4 m) in length and 9 feet (2.7 m) beam, powered by sixteen to twenty oars, one man per oar, all of whom were combatants, there being no separate rowing crew as in the larger ships. They carried a light sailing rig which could be lowered and were shallow hulled, drawing little more than a foot (305 mm) and as such, ideal for working up smaller rivers and tributaries.[7]

Later types

From the late third century AD, types of warship started to change; the upheavals of that period led to a serious diminution of the fleets and of the available manpower, and witnessed the demise of some of the older types. Some recovery of fortune, when it came in the early fourth century, introduced new types of ship. There is a lamentable lack of iconography for the period but enough exists to show that, at least for the northern fleets, a completely new style of warship was in service which demonstrated a total break from the appearance of the earlier types. Mono- and bireme warships with permanently stepped masts and rig and distinguished by prominent bow and stern posts, usually topped by decorative animal heads, were in service. This overall pattern extends to a variety of ships, a memorial from the River Moselle at Neumagen, Germany bears a relief of a river warship, which would have been the successor to the river liburnian; coin evidence shows a similar trend for seagoing warships. The development of ships suited to northern waters, rather than copying Mediterranean types as started at the time of Caesar,[8] by the third century AD had evolved ships of greater beam and draft, stoutly built and with either bireme or double-banked monoreme rowing configurations. To these ships were added lighter, smaller scout ships, here actually designated as a distinct type. These ships were of 'about twenty oars on each side' and camouflaged, their hulls, sails, rigging and even the uniforms of their crews being painted and coloured blue.[9]

The river fleets continued to develop and use boats very similar to the 'Oberstimm' boats, namely open, monoreme vessels of shallow draft. The wrecks of five such boats, dated to the fourth century AD, were discovered at Mainz on the Rhine.[10] Unlike the earlier boats, which had been built 'shell-first' these were built by the 'plank-on-frame' method. The boats are of two types, the first with twenty oars per side and a light sailing rig which could be stowed, was long and slender and clearly a warship; the other was shorter with a broader beam and could be a transport type. The former (warship) was known as a *lusoria*[11] and remained in service until the fall of the Rhine border; on the Danube, it is attested in service in AD 412[12] and continued in service until at least the mid-sixth century AD. The other type could represent the river transport known as a *caudicaria* or *iudicaria* (on the Danube). A further type was in use from the late third century AD, developed specifically for use in the Danube delta, known as a *platypegia*, it was a shallow, flat-bottomed punt with a raised prow and poop, a stern cabin and a light, lateen sailing rig.

The warships which served in the last fleets of the Western Empire in the Mediterranean in the fifth century AD displayed, from the very limited iconography, a further change in design and bore no resemblance to their predecessors. These ships were of modest size (estimated at 110 feet (33.5 m) overall by 16 feet (4.9 m) beam) with perhaps forty oars, double-banked and rowed in a single reme.[13] The ships were decked overall[14] with a single mast and rig permanently erected; they show no towers but could carry some artillery to augment their upturned rams. These ships demonstrated a totally different design philosophy, one that would develop into the *dromon* of the (continuing) Eastern Empire, first referred to as such by Procopius when describing the warships of the Africa expedition of AD 533.

PART III   COMMAND STRUCTURE

6   ORGANISATION OF THE FLEETS

The Republic

There is no indication of any dedicated naval organisation prior to 310 BC. The comparatively small number of Roman warships in service before that date appear to have been operated on a fairly ad hoc basis, being commissioned and directed as and when required.

In addition to their own ships, the Romans could call upon certain of their allied or associated cities or communities to provide ships and their crews. Some coastal areas were required to provide drafts or levies of men for service on Roman ships, but others had to provide fully equipped ships with crews for service under overall Roman command and which could be added to the Roman fleet. This arose from the rather complicated arrangement of treaties and grants of rights of various kinds that the Roman Republic effected with the cities and other states and peoples of the Italian peninsula, as Roman power and influence expanded and brought them into its sphere.[1] Each treaty thus not only defined the relationship between Rome and the other party and also gave certain rights, but, conversely, required that other party to provide men and/or materials for the military. As an example of such an arrangement, Naples, instead of furnishing troops, was bound by treaty in 326 BC, to build, equip, crew and maintain a force of warships to operate in an to protect the Bay of Naples and its surrounding area.

In this manner cities, especially former Italiote Greek cities such as Naples, were rated as *socii navales* and required to provide a quota of warships and transport ships. Coastal towns such as Velia, Heraclea, Thurii, Locri, Croton and Metapontum were so rated. Their contribution to naval forces has been estimated for the early first century BC, at about twenty-five ships[2] and they were instrumental in ferrying Roman forces across to Sicily at the start of the first Punic War in 264 BC. With the progress of this war and the building and deployment of massive fleets by the Romans, the few allied ships that were not lost were enveloped by 'the navy' and a building programme that dealt in hundreds of ships The onus upon the *socii navales* became more for the supply of men, such as in 249 BC, when 10,000 men were drafted to the fleet to replace the losses of the previous year.[3]

Already by the late fourth century BC the arrangements were proving insufficient to provide adequate naval forces, for in 311 BC, two officers, the *duoviri navales*, were appointed as a 'board of admiralty'. This meant that the state had in effect recognised that the navy had become a military service in its own right and in addition to, rather than as an adjunct, of the army. 'The fleet' was regarded as one body at this time, based on the Tyrrhenian coast, although squadrons could be and were detached and sent on missions elsewhere; for example, ten ships were sent to the Gulf of Taranto to support Thurii in 282 BC (with disastrous results). A number of ships were kept on the Adriatic coast once Rome had acquired a coastline there, to guard against the endemic Illyrian pirates. These were not permanent establishments, or designated as detached fleets, but merely forces placed under local command for the duration of the mission, the commander being a senatorial appointment.

This continued to be the case, even with the huge expansion in numbers of ships during the build-up to the start of the First Punic War. During that war, the Senate would vote for a certain number of ships to be provided and commissioned for each campaigning year and appoint commanders; so, for example, just before the war the Senate ordered the building of 100 quinqueremes and twenty triremes, to be added to the existing fleet.[4] Again, in 259 BC, the Romans 'put to sea with a fleet of 330 decked ships'.[5] The war was focused primarily on domination of Sicily and this limited theatre of operations and the enemy's similar concentration of effort, predicated a concentration of Roman naval effort in that area, under direct consular command.

The Second Punic War saw the emergence of what can be recognised as separate fleets, although once again, these were not designated as such, nor probably, intended to be permanent, separate formations. This war ranged far more widely than the Roman's previous wars, with major theatres of operations in Spain and Greece, as well as in Italy and finally, Africa.

Roman plans for the beginning of the war envisaged a retained fleet of 160 ships for Sicily and home waters, while another fleet of sixty ships would escort an expeditionary force to attack Carthaginian territory in Spain.[6] These plans were disrupted by Hannibal's famous march into Italy and his defeating there of Roman armies. The Spanish operation was nevertheless pursued, albeit with a covering fleet reduced to thirty-five quinqueremes, supported by two ships, probably of lighter type, supplied by Rome's ally Marseilles. This fleet, with variations in numbers from time to time, was to remain in Spanish waters until near to the end of the war, when it moved to the North African coast to support the Roman invasion. Thereafter it disappears from the record and must be assumed to have returned to home waters and merged back into the fleet there. In Spain it remained under the overall command of the commander-in-chief there, first Gnaeus Cornelius Scipio and his brother Publius and, after they were killed in 211 BC, by Publius Cornelius Scipio the younger, who delegated operational command to his friend, the capable Gaius Laelius.

The other major deployment away from Italian waters, was a squadron of twenty-five quinqueremes, which was sent in 216 BC to the Adriatic under Marcus Valerius Laevinus (succeeded in 210 BC by Publius Sulpicius Galba, Laevinus moving to command the fleet in Sicily). Thence the squadron progressed into Illyrian and Greek waters in support of local allies, the Aetolians and in opposition to the expansionist ambitions of Philip V of Macedon and especially to thwart his attempted alliance with Hannibal.

These detached formations were only distinguished by the commission of their commander, who was ordered by the Senate to take a certain number of ships for a particular purpose. Although they remained on station for several years, there is no evidence to suggest that any thought materialised that they were considered to be other than temporary formations.

The Later Republic and Civil Wars

The situation remained unchanged into the later third and second centuries BC, when Roman naval attention was drawn increasingly to the eastern Mediterranean. The first substantial change in organisation of the fleet came about with Pompeius' war against the pirates of 67 BC. For this campaign, the Mediterranean and southern Black Seas were divided into thirteen sectors, each under the command of a *legatus*, a rank equivalent to the commander of a legion. Each was allocated a number of ships and men to isolate and attack the pirates in their sector, while Pompeius with a separate force of sixty ships moved to support the legates in turn. The legacy of this was that permanent squadrons were thereafter kept in commission for the Tyrrhenian and Adriatic Seas, with more formations in the eastern Mediterranean. Pompeius' further campaigns in the east led him the set up a naval squadron at Sinope to cover the southern Black Sea and at Ephesus, two squadrons, one for the northern Aegean, the other for the southern part of the Aegean and south coast of Anatolia.

In the period of civil wars of the latter half of the first century BC between Pompeius and Caesar and then between their successors, fleets were raised and deployed by each of the antagonists. They were gathered from whatever commanders and crews could be persuaded to follow a particular cause, which they did evidently from conviction as there were few defections. These fleets included both Roman and to a lesser extent, allied ships, the remnants of the former Hellenistic navies. As an example, the navy of Rhodes joined the fleet of Pompeius in the Adriatic when called upon.[7]

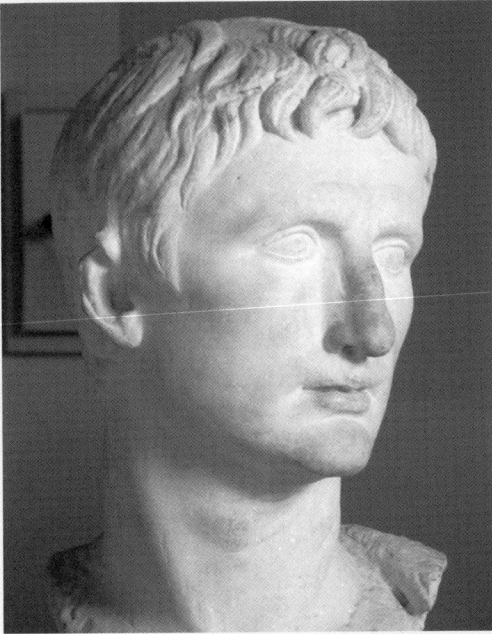

The architects of the Imperial navy. Left, Augustus (63 BC–AD 14). Born Gaius Octavius, he was the great-nephew and adopted son of Julius Caesar, whose name he took: it was on this relationship that he based his claim to power. He is often known as Octavian in this period, until he was awarded the *cognomen* Augustus in 27 BC. His victory at the naval battle of Actium in 31 BC resulted in his being the undisputed master of the Roman world and its first emperor. Archaeological Museum, Istanbul. Right, Marcus Vipsanius Agrippa (63–12 BC). A long-term close friend and ally of Octavian, he was appointed by him to command the fleet, which he did with great ability, building up an efficient and well-founded fleet. With it he first defeated the fleet of Sextus Pompeius and then masterminded the naval campaign that culminated in the battle of Actium. He continued to serve as Augustus' chief minister until his death. Louvre. (*Author's photographs*)

Imperial fleets

At the conclusion of the civil wars in 30 BC, Octavian had acquired fleets of his own and of his former enemies, totalling nearly 700 warships of all types, far more than could be afforded or, in the total absence of any opposition, would be needed. With the start of the Imperial period the whole concept of fleet organisation changed in parallel with the change in its role for the future. Having become a Roman lake, the Mediterranean and, increasingly, the Black Sea, had to be consolidated and policed. To do this, Octavian (now Augustus) and Agrippa established permanent, separate fleets, each with its own identity, commander, headquarters home base and defined area of responsibility.

The system was to be formed around two main *classes* or fleets, based in Italy, with subsidiary fleets at strategic points around the Empire.[8]

The first was the *Classis Misenensis*, based at Misenum at the northern tip of the Bay of Naples. Established by 22 BC, this was to be and remain the senior fleet of the navy and was ranked as praetorian, i.e. part of the emperor's personal guard. The fleet's area of operations was the entire western Mediterranean basin, but it could also (and did) project its power into the Atlantic and established a subsidiary squadron on the Mauretanian (Algerian) coast. This fleet was maintained at a strength in ships and men much greater than was strictly needed to perform its duties. As the senior fleet of the

45

empire, it covered the western Italian coast and transported emperors, members of the imperial family and other notables; it also, importantly, acted as a training centre and a reserve of trained personnel for all branches of the service. These men could be and were sent to supplement other forces throughout the empire and even to provide the manpower for the foundation of other fleets. As an example, men from the Misene fleet were sent to establish the *Classis Britannica* in AD 43.[9]

The Misene fleet remained the principal and strongest of the empire's fleets almost to the end of the Western Empire. It was close enough to be able to be directed from Rome, as relay riders could deliver despatches between Rome and Misenum in one day. It was ideally positioned to be able to project its power across any part of the western Mediterranean basin, as well as to screen the ports of western Italy and the ends of the trade routes to the capital itself. The fleet had local facilities at various other ports and naval stations, for example at Cagliari (Carales), Civitavecchia (Centumcellae) and Aleria in Corsica. Ships were sent further afield from time to time, either for a particular mission or as a temporary detachment, inscriptions of members of the fleet having been found, for example, in Syria and Piraeus.[10] There was a permanent detachment at Ostia and Portus, (when built); another was stationed at Rome, initially quartered in the praetorian barracks, but from the Flavians (later first century AD) until at least the mid-third century AD in their own permanent barracks in the city. There, one of their duties was to attend to the awnings that gave shade to the Colosseum.

The size of this fleet, and indeed that of all of the fleets, is unknown. The names of many ships of the fleet appear on grave stelae and votive altars and Nero was able to enrol a legion (approximately 4,500 men) from among the marines of this fleet in AD 68, later named I Adiutrix by his successor, Galba.[11] At this time, it has been estimated, the fleet had over 10,000 sailors (whether this included marines is not said);[12] at an average of 200 men to crew a trireme, this would indicate a fleet of about fifty

ships. This is, of course, a very crude way of estimating numbers when considering a period of several centuries and a variety of ship types, each with differing crew numbers. Nevertheless, the base itself was the size of a town and the fleet establishment was many thousands of men deploying dozens and dozens of ships for most of its existence.

The second of the Italian fleets was the *Classis Ravennate*, established in about 23 BC at a new base built a short way south of the city of Ravenna, at the upper end of the Adriatic Sea. Slightly smaller than the Misene fleet, it was also rated as praetorian and had as its area of responsibility the Adriatic and Ionian Seas and, being adjacent to the mouth of the River Po (Padus), the navigation of that river system. This enabled the fleet to be a part of the protection of Italy north of the Apennines. The fleet could and did also operate around the Peloponnese and into the eastern Mediterranean.

Like the Misene fleet, a detachment from this fleet was stationed in Rome, again with their own quarters. The fleet's harbour was one of the best on the Italian Adriatic coast, which has few natural harbours and the location also linked with the northern end of the Via Flaminia, a direct link to Rome. From their location, the fleet could provide rapid connections and communications with the north end of the Adriatic (through the port of Aquileia), to the eastern Alpine and upper Danube regions, or across to Split (Salonae), the Dalmatian coast (previously, with its myriad islands, a notorious haunt of pirates) and connections with the middle Danube area. In the south, stations were maintained at Ancona and Brindisi, the latter one terminal of the route to Durres in Albania (Durazzo, Dyrrhachium) connecting with the Via Egnatia through the Balkans to Thessaloniki (Salonica) and Byzantium (later Constantinople/Istanbul). There were two other stations in western Greece, to cover the Gulfs of Patras and Corinth and the Ionian Islands and passage along the western Peloponnese. Units of the fleet operated from time to time in support of the Misene fleet, especially in the third century AD, with frequent campaigns in the East.[13]

Provincial fleets

Two other fleets were established for the eastern Mediterranean, the *Classis Alexandrina* and the *Classis Syriaca*. The former was to control the sometimes troublesome North African and Palestinian coasts and to oversee the increasingly important trade route, including grain transports, from Egypt to the West. The policing and regulation of traffic on the River Nile was the responsibility of the *potamophylacia*, a separate river police force organised by and inherited from the Ptolemies that had its own men, ships and bases on the river. This could be, and was from time to time, augmented by the fleet when needed; the *potamophylacia* was wholly absorbed by the fleet in the second century AD.[14]

One other area of operations for this fleet was the Red Sea. The Romans did not maintain a permanent fleet on this sea, but did organise a fleet in 26 BC with personnel drawn from the Alexandrine fleet for a military expedition to what is now Yemen. This fleet had eighty warships and 130 transports, the latter being requisitioned local merchant ships. The warships, which in the absence of any anticipated opposition (there was in fact none) need only to have been of the smallest types, were 'built' on the Red Sea shore, presumably from prefabricated parts brought overland. Some were provided by the allied kings of Nabatea and Judaea, who also contributed forces for the venture. It is possible that the Nile–Red Sea canal was in use and that some of the ships could have been brought from the Mediterranean by this route. The canal was prone to silting if not constantly maintained; Trajan (ruled AD 98–117) restored the canal and with his annexation of the kingdom of Nabatea in AD 106, had both sides of the northern Red Sea under Roman control. Even so, there remains no evidence for any but occasional forays by the Alexandrine fleet on to the Red Sea.

The *Classis Syriaca* had its headquarters at Seleucia near Antioch on the north Syrian coast and was placed to cover the Levantine coast as well as the south coast of Asia Minor, also previously a notorious centre of piracy. The fleet also extended into the southern Aegean Sea and being the closest to the ever-present threat of Parthian and Persian power to the east, was instrumental in maintaining transport and communications links with the West, frequently having to transport troops to oppose threats or attacks from the east.

Although each fleet was a totally independent entity, their spheres of responsibility could and did overlap. Ships from separate fleets operated together seamlessly for differing operations, ships from other fleets being drafted to assist in major operations, such as the transport of troops and supplies for campaigns against the Parthians, or Trajan's campaigns across the Danube.

The North African littoral of what is now Algeria and Morocco had been in a state of unrest and occasionally open revolt after the emperor Gaius (Caligula, emperor AD 37–41) had its ruler murdered. Under his successor, Claudius (emperor AD 41–54), the whole territory was brought under direct Roman rule in AD 41 and 42 and formed into the provinces of Mauretania Caesariensis (the eastern part) and Mauretania Tingitana (the western part). These campaigns were supported by the *Classis Misenensis*, augmented by ships of the Alexandrian and Syrian fleets. The capital of Caesarea (Caesariensis) received a naval base with its own harbour, distinct from the merchant harbour and which became home to a permanent naval detachment or squadron. This unit was made up from ships and men from the Alexandrine fleet, but was not constituted as a fleet in its own right, but remained an adjunct of its parent fleet, which was well able to supply the required ships and men as well as support, from the relative tranquillity of the eastern Mediterranean.

The squadron, although sufficient to patrol the coasts, including the Atlantic seaboard,, was not able to deal with major conflagrations and Misene ships had to intervene to suppress raiding by Mauretanians in AD 170 and 171. It intervened again in AD 260, to help suppress revolts in Africa and Numidia (part of modern Algeria). There were more peaceful interventions when naval personnel were employed to apply their abilities in civil works, for example, in AD 152, the engineer in charge of building an aqueduct at Saldae in Mauretania reported that 'the constructor and his workmen began excavation in their presence, with the help of two gangs of experienced veterans, namely a detachment of marine infantry and a detachment of alpine troops . . .'[15]

The expansion of the empire to the line of the Danube under Augustus, completed by 12 BC, engendered the formation of two more fleets for the defence of that river. The Danube was naturally divided into upper and lower parts by the Iron Gates Gorge (between Orsova and Donti Milenovac, about 100 miles (160 km) east of Belgrade), which was at that time an impassable torrent. For the new border adjacent to the provinces of Noricum, Rhaetia and Pannonia (approximately modern Switzerland, Austria and western Hungary), flotillas previously formed and used in the advance on the rivers Sava (Savus) and Drava (Dravus), were moved up to the Danube and reinforced to form the *Classis Pannonica*, with headquarters at Zamun, near Belgrade (Taurunum).

For the lower Danube, from the Iron Gates to the Black Sea, the *Classis Moesica* was formed, with headquarters at Isaccea (Noviodunum) in Romania. This location is nearer to the Danube delta than the centre of the fleet's stretch of the river because this fleet's responsibilities extended into the western Black Sea and the sea routes between the river mouth and the Bosporus. With the increase of Roman power into and around the Black Sea area, the fleet acquired the further duties of protecting Roman interests in the western half of the Black Sea, including the Hellenistic cities around its northern shore and the strategic link with the Bosporan kingdom of the Crimea and adjacent lands, an important producer of grain. The fleet later established a naval base at Chersonesus.

To cover the southern shores of the Black Sea and to maintain a watch over its east coasts, the *Classis Pontica* was formed. The Romans had had some forces on the north coast of Asia Minor since they were organised by Pompeius

in the sixties BC. The last client king of Eastern Pontus, Polemo II, was 'retired' in AD 63 and his kingdom annexed and made part of the province of Galatia. The former royal fleet was taken over and merged with the Roman ships to form the fleet, with its headquarters at Trabzon (Trapezus), later moved to Cyzicus (on the Sea of Marmara). For once there is a little evidence of the strength of this fleet, which was noted at some forty ships.[16] This fleet extended its influence into the south-eastern Black Sea to eradicate the spasmodic piracy there and to cover the important military supply route from the Bosporus and Danube Delta, to north-east Asia Minor, to supply Roman forces facing Armenia and the Parthians. Subsidiary bases were set up in the second century AD near Poti (Phasis) in Georgia and Asparis, near the present-day Turkish–Georgian border.

The other great riverine frontier of the empire, the Rhine, also had its own fleet, the *Classis Germanica*. Caesar had established the border of the Roman empire on the Rhine in the mid-first century BC, leaving garrisons with some small craft for patrolling, at intervals along its length. Legionary bases grew into towns at, for example, Mainz, Koblenz and Xanten. In 12 BC Augustus resolved to advance the border to the River Elbe (Albis) and in that year the future emperor Tiberius and his brother Drusus crossed the Rhine. Drusus concentrated the Roman's existing ships and boats at Bonn and formally constituted them as the *Classis Germanica*, with its own *praefectus* and administration, as for the other fleets. He also had new ships built and brought trained crews from the Italian fleets by way of reinforcement.[17] Fleet headquarters was at Cologne (Colonia Agrippinensis) and apart from the Rhine and Moselle, it was charged with patrolling and incursions into the tributaries entering from the right bank, such as the Neckar, Main and Lippe. Seagoing ships had to be acquired to cover the mouth of the Rhine and adjacent coasts. Later, with the addition of Britain to the empire, it had to operate jointly with the *Classis Britannica* to maintain the essential link between the armies of the Rhine and Britain. After the

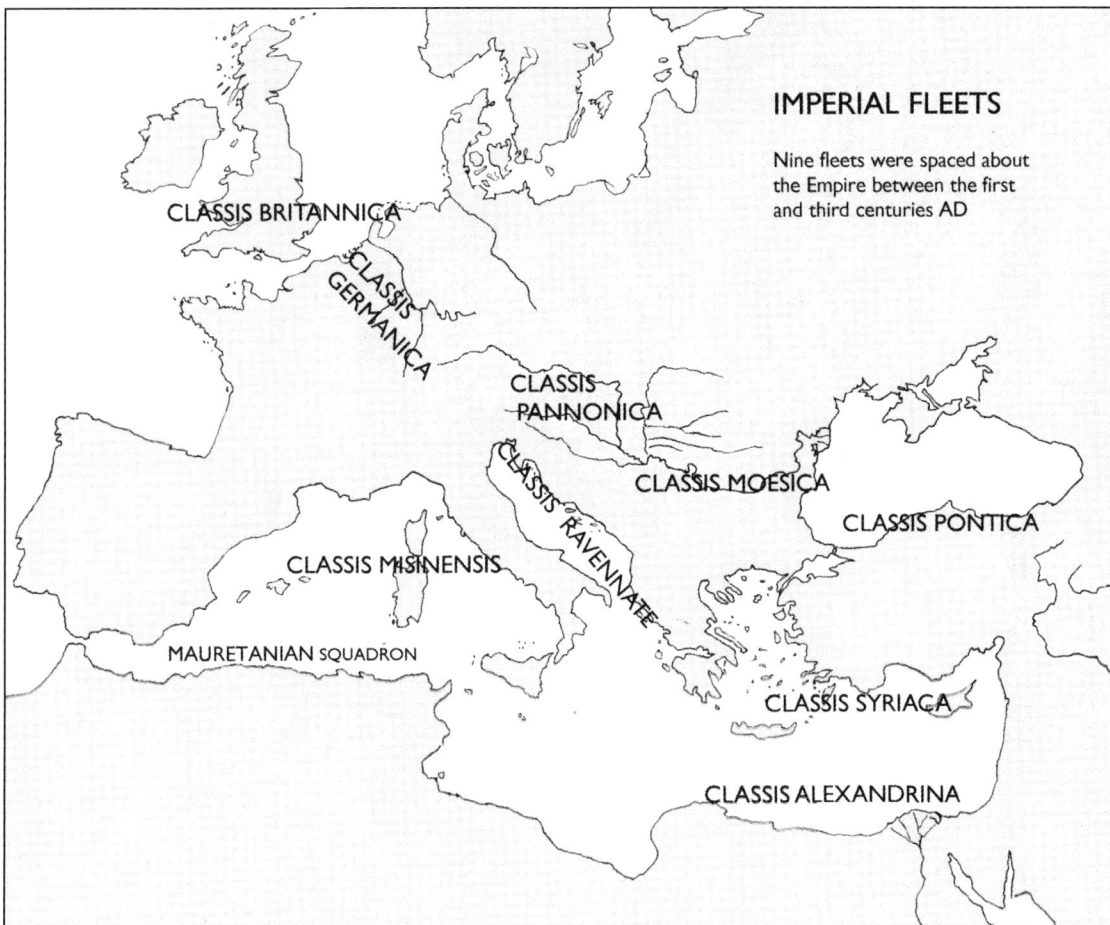

IMPERIAL FLEETS

Nine fleets were spaced about the Empire between the first and third centuries AD

CLASSIS BRITANNICA

CLASSIS GERMANICA

CLASSIS PANNONICA

CLASSIS MOESICA

CLASSIS PONTICA

CLASSIS RAVENNATE

CLASSIS MISENENSIS

MAURETANIAN SQUADRON

CLASSIS SYRIACA

CLASSIS ALEXANDRINA

loss of the embryo province east of the Rhine after AD 9, the river became the permanent border, later altered from the middle reaches up by the advance into the Agri Decumates (the re-entrant between the Rhine and Danube) between the late first and third centuries AD.

After his conquest, Caesar had left ships on the north coasts of Gaul to patrol, deter any piracy and secure the trade in the English Channel. A few such ships under local military control had been sufficient for the predominantly peaceful area but in the forties AD Claudius resolved to add Britain to the empire. The ships were formally redesignated as the *Classis Britannica* and reinforced by new building and by ships with crews and specialist personnel brought around from the Mediterranean by sea. Preparations for the invasion had most likely, in

view of their extent, started in the reign of Gaius, but were completed by his successor, whose forces invaded in AD 43. The new fleet was augmented by ships from the *Classis Germanica* and was vital to the success of the invasion which depended wholly on supplies from Gaul. After the initial invasion, the *Classis Britannica* had to continue to grow and extend its area of operations, as the Romans expanded their area of occupation, the fleet eventually operating right around the British Isles. The fleet's prime purpose would remain however, to maintain the essential links with the mouth of the Rhine and the armies there, as well as with Gaul and the fleet headquarters was accordingly established at Boulogne and with another, a little later, at Dover (Dubris).

There were two other formations classified

as fleets which do not seem to have been permanent, but pass briefly through the surviving records, indicating that they were formed for a particular purpose, at the ending of which they were disbanded. The first was the *Classis Perinthia*, formed by Claudius in AD 46 to cover his annexation of Thrace, after which there is no other indication of its continued existence.[18] Thereafter responsibility for the Thracian coast passed to the *Classis Moesica*.

The other formation was the *Classis Nova Libyca*, which appears from the scant references, to have been formed in the late second century AD to reinforce the Libyan shores at a time of unrest there. It is not heard of again beyond the mid-third century AD and either was disbanded, it's ships returned to their parent fleets, or was lost in the great upheavals of that time.[19]

The late Empire

These nine imperial fleets continued to operate for nearly 200 years until caught up in the upheavals of the third century AD. With the Empire hard pressed by internal dissention and external pressure, the fleets could not be maintained as before, as the first call on available manpower, resources and money was the army. Neglected and denied resources, the great praetorian fleets deteriorated to a shadow of their former selves, as did the other Mediterranean fleets. The Black Sea was progressively abandoned and the riverine fleets seriously overstretched and at times overwhelmed.

With the accession as sole emperor of Diocletian in AD 285, stability was returned to the empire, together with the need to reorganise the remains of the fleets that he had inherited. The *Classis Britannica*, due to its particular location and function as a mainstay of the province's garrison and defence and in the face of increasing barbarian seafaring activity and ability, had remained the least neglected and probably the best fleet left in the empire. It was part of the command of Carausius, commander in Britain, who improved its strength and efficiency and went on to the offensive against the sea raiders. It alone continued as recognisably the *classis* of yore.

The rest of the fleets in the Mediterranean were reorganised into a greater number of smaller squadrons, rather than try to reconstruct the great fleets. Each squadron was commanded by a *praefectus* and assigned to a military district and placed under the overall command of the military commander-in-chief for that district; in so doing the fleets lost their former independent identities as *classes*. Thirteen such squadrons were formed and became the basis of naval organisation in the Mediterranean thereafter.

The Danube border was reorganised into four new provinces, Moesia Prima and Secunda, Scythia and Dacia Ripensis. The former *Classis Moesica* was also divided into four parts, one allotted to each of the new provinces and again, under overall command of the local military commander. The fleet remained independent only insofar as that part of it based in the Danube delta and responsible for the Delta and with the Thracian coast. The *Classis Pannonica* disappears from the record in the third century AD, but Roman naval forces on the Upper Danube are known from the fourth century AD, once again as units forming part of the border forces, rather than as an independent fleet as before.

The *Classis Germanica* had all but ceased to exist by the mid-third century AD, local commanders having to employ whatever ships and crews they could acquire on an ad hoc basis. On the restoration of the Rhine frontier by the co-emperor Maximian after AD 286, naval forces were again built up but again on an area by area basis and allocated to local military commanders, the *classis* was not reconstituted as such.

These dispositions continued to serve through to the late fourth century AD, when with the loss of territory, including increasing parts of the Mediterranean seaboard, even the squadrons lessened in numbers, especially in the west until Roman naval forces were formed from whatever ships could be amassed and crewed.

Paying for the fleets

The Roman state, like most others, gathered funds by way of taxation and of tribute or levies from allied and federated states. Out of this, the Senate, as the governing body of the Republic,

Warships were an enduring theme on the reverse of Roman coins. Top left, copy of a Republican bronze aes grave, mid-third century BC, showing a warship prow, probably a heavy warship; top right, a sestertius of Commodus, late second century AD. Below left, a bronze denarius of M. Agrippa with a winged victory figure above the ship prow, late first century BC; right, gold aureus of Hadrian, showing the emperor's ship, AD 119. (*Author's photographs*)

voted funds for the raising and maintenance of armed forces. This function was replaced in due turn by the imperial administration that succeeded the Republic.

There was no regular, annual budget and voting of funds under the Republic; instead a call to arms and requisite funding was provided as and when circumstances required. In the state of almost continual warfare in which the Republic found itself, this could be and often was needed, more than once a year. Provision could be made for all-out war, but also for specific, more limited purposes, either as part of a larger conflict or in isolation.[20] Such processes included naval forces but after 311 BC and the appointment of the *duoviri navales*, annual funds had to be allotted to pay for this permanent administration and establishment of the navy.

It was once more the growth for and during the Punic Wars that multiplied the need for and amount of regular naval funding. The strain of funding these wars, so much larger for Rome than any that had preceded them, exhausted their treasury, being more than could be covered by normal sources of income. In 241 BC to fund the fleet that gained the final, decisive victory of the First Punic War, public subscription was sought and the richest of the citizenry, singly or in syndicates, made loans to the State to build and fit out ships for the fleet. The loans were repayable by the State upon victory and the confidence of the lenders was rewarded.[21]

A similar lack of funds during the Second Punic War in 214 BC was dealt with by the Senate making a levy to fund the fleet. In that year, Livy reports that

The consuls, on the Senate's authority, issued an edict to the effect that anyone whose property, or whose father's property had been assessed . . . between 50,000 and 100,000 asses or whose assets had subsequently increased to that figure should furnish one sailor with six months' pay; those assessed at over 100,000 asses and up to 300,000, were to furnish three sailors with a year's pay; those whose assets were between 300,000 and one million, were to make themselves responsible for five sailors and those with more than a million, for

the Roman's previous experience, demanded more diversity in command structure and from 264 BC, the Senate could appoint a *praetor* to command a detached squadron; this was a senior rank, equivalent to that of a governor of a province, for which a man of the senatorial class was required, unlike the tribuneship which drew its men from the lesser equestrian class.[2] These commissions would only last until the end of the particular mission for which it had been granted.

This command structure worked well throughout the First Punic War, operational command of the fleets at sea remaining with the consuls and their tribunes. For the wider-ranging Second Punic War, in addition to the main fleets in Italy, other fleets were sent for service in Spain and Greece. As before, command of these detached forces was entrusted a *praetor*, such as Marcus Valerius Laevinus, who was sent in 214 BC, with twenty-five ships to Brindisi to assume command of the forces there and support Rome's allies in the First Macedonian War.[3]

This system remained unchanged until the war against the pirates waged by Pompeius in 67 BC. Pompeius divided his intended area of operations, in effect the whole Mediterranean, into thirteen sectors and appointed a *legatus* to each as its overall military commander for sea and land forces. A *legatus* (legate) was the rank to command a legion. Eleven more were appointed for the forces Pompeius had with him and in addition, he appointed a *praefectus classis et orae maritimae* (prefect of the fleet and the coast).[4] Nothing further is heard of the *duoviri* and *praefecti* and perhaps their office and functions were merged with or subordinated to that of the new *praefectus*. This would seem unlikely as the newly created appointments are not mutually exclusive. The existing administration system had matured and worked well and it was the neglect and consequent weakness of the navy that had permitted the growth of piracy, rather than any failing in its organisation. As the navy continued to function after the war against the pirates and no more is heard of Pompeius' special appointments, it must be concluded that the extra appointments were provided for that purpose only and that the former organisation continued afterwards. While at the close of these operations, surplus ships were laid up and the extra forces recruited for them were stood down, permanent squadrons were kept in commission and posted to various areas, each continuing to be commanded by a legate.

Imperial high command

After 30 BC a new high command structure was adopted reflecting the division of the service into various permanent fleets. There was no centralised navy high command although a staff officer from each of the Italian fleets was with the military staff attending the emperor, together with a staff officer from any of the other fleets in whose area the emperor might be. All responsibility for command of each fleet and everything to do with it was vested in the *praefectus classis* (fleet prefect), who was appointed by and answered to the emperor; the *duoviri navales* were dispensed with. Given the distances and the time that communications took at that time, this was the only practical method for the command of widely separated forces. The command of each of the two Italian fleets was entrusted to a *praefectus* appointed by the emperor and of at least equestrian rank and rated equivalent to Praetorians, i.e. guards; these posts would be held by ex-legionary tribunes and under Claudius by imperial freedmen. The consular and senatorial appointments ceased. Provincial fleet commanders, again appointed by the emperor, but under the local overall command of the provincial governor, were now also classified as *praefectus classis*. Under the *praefectus* was his subprefect as executive officer and aide-de-camp, a *cornicularius* as next-ranking officer and a number of functionaries who were probably leading rankers rather than officers, of various types, called *beneficiarii* (appointees), *actuarii* (clerks), *scribae* (writers) and *dupliciarii* (this last as senior or leading ratings, on double pay), all of whom made up the Prefect's administrative or office staff. One position from Republican times that was continued into the Imperial navy was that of *quaestores classici* for each fleet: a *quaestor* was

This statue of the first century AD has a cuirass embossed with sea nymphs and dolphins, which would suggest that the statue was of a senior naval officer, a *navarch* or even a *praefectus*. The overlapping scales beneath the cuirass are embossed with lion and elephant heads. Thessaloniki Archaeological Museum. (*Author's photograph*)

primarily a financial officer, hence these were fleet treasurers.[5]

Claudius replaced the fleet prefects, traditionally of equestrian or even senatorial rank, with appointees who were professional bureaucrats, with *procurators* who were usually imperial freedmen, i.e. ex-imperial slaves and non-military men. Vespasian (emperor AD 69–79) reverted to the appointment of experienced military men. The Italian fleet prefects were now second in rank only to the Praetorian prefect, and once more of at least equestrian class.

These command structures continued through the time of chaos in the third century AD, when much of the navy itself ceased to exist. The naval forces that survived continued to be commanded by a *praefectus* for each of the squadrons that succeeded, each based on a

local centre under the overall command of the *dux*, a new rank for the overall district military commanding officer. There were thirteen such squadrons but no indication that social rank was a continuing requirement for senior officers. A further rank was added from the third century AD, namely that of *praepositus reliquationis*; this was a temporary flag rank for an officer commanding a detachment or in the absence of the *praefectus*[6] rather in the manner of the old republican appointment of a *praetor* for a similar purpose.

The emperor Constantine greatly reformed the armed forces, in most cases appointing local area commanders who had both troops and fleets under them and with no separate naval command, high or otherwise. Senior military officers in the late period of the Empire were rated as *magister militum* or *magister equitum* (for cavalry) but there is no record of a *magister classis* and fleets continued to be commanded by *praefecti*, who were again subordinate to their local *duces* or military commanders. The exception was probably the British Fleet which retained more of its old, early empire structure, due to its particular area of operations in that the 'frontiers' of the province were of course, mostly sea. In AD 364 there is a mention of a *comes maritimae* in Britain, a supreme naval commander, but this is the sole mention of this post and it may well be a one-off commission for a particular purpose.

For the remaining life of the Imperial navy it seems that military commanders were appointed largely on a provincial basis to include whatever naval forces fell under their sway, there being no mention of any separate ranks that refer to naval forces alone. Roman naval squadrons enjoyed a brief resurgence in the early fifth century AD, before being finally lost in the abortive attack on Vandal-held Carthage in AD 467.[7] No details survive of the organisation of these last Roman fleets, the command of which resided with the military commander of land forces, appointed by the emperors and there is no record of a strictly naval command structure, beyond the captains of individual ships.

The crews

The Romans were able to and did fit their well-established 'century' system of military administration to the organisation of ship's crews, a system which lent itself to the purpose. Although the census divided the Roman manpower into bodies of a hundred, once the halt, lame, insane or plain undesirables were removed, the century of fit men for military purposes numbered an average of between sixty and eighty men and which became the basic military unit.[1] The 'century', originally of sixty men, was increased to eighty men by the mid-second century BC, at which level it remained. A centurion was appointed to command the century, who nominated his *optio*[2] and two or three others, a practice carried over into imperial times.

The crew of a ship was classified as a naval century, convenient enough for billeting them ashore, but which is complicated when trying to fit the system into the crewing requirement of different types of ship and which would not necessarily, conveniently fit such numerical divisions. There were fundamental distinctions between the branches of a crew that made up the complement of a Roman warship, and so ashore, naval personnel would have been organised into centuries of rowers, marines, sailors and trades respectively, each requiring a different training regime. The ships' officer corps was a separate entity in that an officer ordered to captain a warship needed to appoint his officers, his first lieutenant or *gubernator*, second lieutenant or *proretus* and rowing master or *pausarius*). To form his crew, he would then have to draw the requisite number of rowers from a century of *remiges*, with their own officer and leading rates, the appropriate contingent of marines from a marine century with their own (military) officer and the necessary number of sailors and tradesmen, such as a carpenter and medical orderly from other, specialist centuries.

The actual method of crew allocation from republican times is not known and can only be reconstructed from the application of known military organisation.

The actual crew requirement of a fifth century BC trireme is known from the records that have survived for the Athenian navy.[3] The crew totalled a little over 200 men: four or five officers, sixteen ratings, 170 oarsmen and a marine company of anything from fourteen to forty men. In detail, there was a captain, a first lieutenant, a rowing officer and a bow officer (second lieutenant); of the ratings there were two helmsmen, two stroke oarsmen, a carpenter, a flautist or musician and up to ten sailors. The rowers were divided as to fifty-four thalamites (the lowest level), fifty-four zygites (the men of the middle level) and sixty-two thranites (the topmost level); these last men were the senior hands who controlled the two other below them in their group, as they were the only ones who could see out and direct the synchronisation of their collective oar-stroke. Of the marines, the Athenians, relying mostly on ramming tactics, kept their numbers to the minimum, carrying only ten marines and four archers. Other Greek navies carried up to forty marines and we can assume that the Romans were in that latter category, so as to make the best use of their superb infantry. These numbers varied as the form of the trireme varied over its long service history, but the exact extent by which they did so is not now known; it could be reasoned that as the marine complement of a quinquereme, a much larger ship, in peacetime was forty men[4] or a half century and therefore that the normal peacetime marine unit of a trireme would have been some twenty or so men, a quarter century and thus appropriate to be commanded, in Roman service, by a *suboptio*.

Ranks and terminology

In the beginning, due to their very strong

influence at Rome, it is most likely that predominantly Etruscan nautical terms were in use, particulars of which have disappeared along with the understanding of their language. This influence could well have provided some of the terms and expressions that are considered to be Roman, i.e. Latin and which are in fact, of Etruscan origin.[5] There were also some Carthaginian influences (Rome was allied to them for over two centuries). After the progressive Roman expansion southward into and absorption of the Italiote Greek areas of southern Italy, some Greek names for officers and shipping terminology started to be adopted and adapted by the Romans. Thereafter, an amalgam of Greek and Latin nautical terminology evolved and continued in general use throughout the naval service for the remainder of its existence.

The senior seagoing rank was a *praetor*, a rank either augmented by or succeeded by that of a *legatus*. These ranks would equate in their various seniorities to modern commodores, up to admirals. There were also military tribunes, although these would appear to be senior marine officers, rather than ship commanders. The captain of an individual ship was a *trierarchos*. Originally master of a trireme, the term had come to be applied to the commander of any type of warship. It was a rank rather than an appointment and the trierarch could also command shore detachments; appointment was usually by promotion from the lower ranks. Ironically, the term 'captain' that we now use is from the Latin for 'head man'. The Latin equivalent was *magister navis*, ship's master but this was more commonly used for a merchant ship captain. The distinctions were not so clear-cut in that whereas a trierarch might command a trireme or larger type, he could also command a smaller one. Ships could also be commanded by lesser ranks, thus a junior officer, although only an *optio*, i.e. second in command to a centurion, could be the captain or trierarch of a liburnian[7] in the same way that today, a lieutenant put in command of a ship is that ship's captain.

Subordinate officers on board were the *gubernator*, 'first lieutenant' or navigating officer, in charge of the steering and after part of the ship,

second in command to the trierarch; next in seniority came the *proreus* or *proretus*, the bow officer in charge of lookouts, depth sounding and the forward part of the ship. The rowing officer was called the *pausarius* or *celeusta*. All of the names of these ranks were inherited from Greek practice and Latinised. The Roman names for the three classes of rowers (thranites, zygites and thalamites) are not known.

Imperial ranks

With the organisation into fixed fleets with permanent administrative staffs to organise crew allocation, there is no evidence to suggest any drastic reorganisation of personnel, rather that the well-tried usages of the past were simply carried over and applied for each of the fleets, with crew allocation as a function of fleet headquarters. Adopting another Greek word, the senior seagoing rank became a *navarchos*, originally a ship master, but which by the late fifth century BC had become the term among the Greeks for a squadron commander or admiral. It first appears in Roman service with the Imperial navy and was the highest rank attainable by a non-Roman citizen until the reign of Antoninus Pius (ruled AD 137–161). A senior admiral was rated *navarchos princeps* (chief navarch, an amalgam of Greek and Latin terms) and this seems to have superseded the former Republican ranks of *praetor*, *tribune* and *legatus*.

Ship officers continued to be the trierarch, *gubernator*, first lieutenant, *proreus*, second lieutenant, and *pausarius*. Under-officers were a *secutor* or master-at-arms, responsible for discipline; the term *nauphylax* (Greek, ship guard), may well have been for an officer of the watch, i.e. a duty, rather than an actual rank. Larger crews and shore bases would have a physician and include *faber navalis* (naval craftsman, i.e. shipwright). There were a number of other senior ratings perhaps equivalent to petty officers and every crew would include a *medicus* (medical orderly) and a ship's carpenter (*faber*). Depending on the size of ship, a number of other men were added to the crew and called either a *beneficiarius*, *scriba* or a *librarius*, the ship's clerks or pursers. In addition, some of

Grave stelae showing naval careers. Left, Lucius Octavius Elattes, described as a *gubernator*, or first lieutenant and navigating officer, who died aged fifty-eight years, having served in the navy for thirty-four years. Mid-first century AD. Romisches Germanisches Museum, Cologne. Right, Horus Pabecus from Alexandria, a *proretus* or bow officer and second lieutenant, a naval veteran who had served with the German fleet and died there, aged sixty. Romisches Germanisches Museum, Cologne.

Above, Lucius Valerius Verecundus, a marine of the *Classis Germanica*. He is from the fleet's first cohort, the century of Ingenius. He died aged only twenty-five, after four years service. Romisches Germanisches Museum, Cologne. Right, Aulus Platorius Sergius, who was legatus or legionary commander of Legio I Adiutrix. Nero had a good relationship with and was popular with the fleet and in AD 65, when he started to fear for his safety, he formed this legion from personnel of the fleet at Misenum. Archaeological Museum, Aquileia. (*Author's photographs*)

these and of the other ratings were classed as *dupliciarii* or 'on double pay' and were what we would now class as leading seamen. There are other specialist ratings known from the imperial navy, namely a *symphoniacus* (musician), the equivalent to the Greek *auletes* (flautist). There was also a *bucinator* and a *cornicen* (players of the *bucina* and *cornu* respectively), metal wind instruments giving distinctive sounds, the equivalent of bugle calls. A navy diver was called a *urinator* (from the verb then meaning to dive), and a *coronarius* was responsible for dressing the ship with wreaths and bunting for festive or ceremonial occasions; such a speciality duty did not justify a rank and these men were normally signallers. There was also recorded a *dolator* or *dolabrius* from an inscription in Germany, who seems to have been a pioneer or engineer but whether this was, from the very limited source, a rank or a duty allotted, is not clear.

The ship crew included a number of sailors for ship-handling, called a *velarius* and being named after the *velum* or sail. The rower was a *remex*, after *remus*, an oar. There was a *pitulus* or *pausarius* (time-keeper) who would beat the oar timing with a mallet, *portisculus*, additionally the stroke oarsmen, who set the pace and the stroke and who were the two rearmost. Depending upon the size of the crew a number of the other rowers would be *duplicarii* or leading hands. It is not known how many men there were to each leading hand but in the army the smallest unit was a *contubernium* of eight men, one of whom was senior and a figure of about eight to a dozen rowers per leading hand would fit well enough. All ranks, rowers, sailors, as well as marines, were rated as *miles* or soldiers, as opposed to those of the merchant ships, who were classified as *nautes* or sailors. Several of the navy's sailors and marines described themselves as *manipularii*, or men of a maniple which was a military formation of two centuries, the equivalent perhaps of the crew of one of the larger ships, of which these men could have been part. The continued use of this term in the naval context is notable; manipular tactics, whereby centuries were joined for battlefield manoeuvre, were introduced in the legions in the early fourth century BC, but went out of use before the end of the second century BC.[8] By the time that members of the imperial navy described themselves as *manipularii*, the term was long obsolete, lending credence to the above interpretation.

For the military part of the ship's complement, marines, archers and artillerymen, the command structure followed that of the army. The fleet thus had centurions of varying grades, just as in the army's centurionate and being experienced and competent men who had risen to the rank, as opposed to the Fleet Prefects who were appointed and might be amateurs, as happened from time to time. Fleet centurions commanded marine units aboard ship with a seniority equivalent to the size of ship and the size of its detachment of marines. For a larger unit or a shore posting, an *optio* or second-in-command would be appointed by the centurion and has been seen above, the rank could stand alone as a command rank. Larger ships would also carry an *armorum custos*, an armourer in charge of weapons and armour; finally there was the *suboptio*, a rank unique to the navy. It would seem likely that smaller units of marines, of a dozen or so, for the small ships, or even a small ship, would have been sufficiently commanded by an *optio* and *suboptio* alone. Marines would sometimes designate themselves as *milites classiarius* or 'fleet soldier' on grave stelae, presumably to differentiate them from the *miles*, who could also be sailors.[9]

The hierarchy of command between the 'seamen' and 'military' functions is unknown and in the close confines of a ship there must have been a certain degree of merging of the functions; nevertheless the structure suggests, at least in the larger ships, that the 'sailor' officers held supreme command of the ship, with the 'military' officers taking over when about to go into action or for shore duties. Naval and marine personnel were on occasion, posted ashore and classified as vexillations, for various duties. There are instances of them assisting the *vigiles* (a type of police and fire service) in Rome and working on an aqueduct in Africa,[10] as well as assisting in the building of the shore fort system in Britain and Gaul.

PART IV   THE CREWS

9   RECRUITMENT

Manpower

To fill the ranks of its armed forces, the early Roman state relied upon a citizen levy of all able-bodied men between the ages of seventeen and sixty who were liable for military service. A census was taken of all male Roman citizens, who were organised into 'hundreds' or centuries, according to relative wealth. As each man was responsible for providing his own weapons and equipment these classifications determined the type of troops that each century would become, ranging from the richest who could afford a horse and the trappings for it as well as their personal equipment to become the cavalry, via descending orders of wealthiness, each providing various levels of helmet, armour and personal weapons for varying classes of infantry from heavy to light, to the poorest who could become slingers.[1] There were men left over from this, too poor or unsuited for any of the above, called *capitecensi* (literally 'head count') but who were nevertheless useful for a miscellany of support tasks, such as bearers, handling supply wagons and animals, manning field kitchens, mending equipment, repairing defences or tending wounded and any other task that could be found.[2] It is more than likely that it was from these *capitecensi* that the first rowing crews were drafted, no equipment being needed to man an oar, the oar being part of the ship's inventory. Sailors could be levied from Roman merchant ship crews and some of the lighter infantry and archers could be assigned, to provide the fighting component of a complete warship crew (the heavily armed infantry would always be needed for the army and were in any event, less suitable or wasted aboard a ship). It should be emphasised that there is no hard evidence to support the above contention but, faced with the fact that ships were manned and given the known framework of Roman organisation, this is the most likely method adopted and one that fits exactly into that organisation system without variation of it.

On successive mobilisations of the armed forces, men would go to the same centuries to which they had been assigned initially and those centuries of men were habitually called or assigned to a particular maniple, cohort or legion. It is reasonable to assume that the formations or units outlined above would go to their own respective ships or naval postings, the men gaining in experience with each succeeding campaigning season and becoming in effect if not in fact, naval units or centuries.

Whether by alliance, absorption or conquest, the expansion of Roman power and hegemony across Italy was to make fresh sources of manpower for the navy available as the Romans encompassed neighbouring peoples and especially maritime peoples. Although not Roman citizens (who were liable for service in the legions and as mentioned above) these allies were liable to provide contingents or drafts of men for military service, being formed into units similar in form to those of the Romans and under overall Roman command, but classified as *auxilia*. In connection with the navy, of especial note is that certain coastal cities and communities, such as Ostia and Anzio, were specifically exempted from this military levy[3] but instead, had to furnish and fit out ships for service and to provide crews to man them. As an example of this, in 326 BC, Rome entered an alliance with Naples, a term of which was that the city was not liable to provide troops, but instead had to operate enough suitable ships to patrol and guard the harbour, effectively the whole, strategic, Bay of Naples area.[4]

From about 310 BC in the period preceding the First Punic War, which was to be above all a naval war, the military rosters detailing contingents to be supplied by federated and allied peoples for the army, omit the maritime Italiote

Greek communities. These communities must have been committed instead to provide a large proportion of the greatly increased manpower that was required and in fact mobilised and served in the navy for the Punic Wars.

Roman expansion in the peninsula from the sixth and fourth centuries BC, first northwards, brought to them the long-established seafaring traditions and abilities of the Etruscans. Following on this, their further expansion southward into and beyond the Bay of Naples, brought to them the Italiote Greeks; with them came the accumulated experience of Greek seafaring and shipbuilding, as well as a further large supply of seamen.

It must be emphasised that, contrary to some popular belief, the Roman navy did not use slaves at any time to man the oars of its ships; the concept of the 'galley slave' chained to his oar was an invention of the sixteenth to eighteenth centuries AD and unknown in the ancient world. The crews of Roman warships were free men from the citizen levy, allied contingents supplied by treaty and volunteers. There are in fact, only two instances attested and no doubt remarked upon because of their uniqueness, when slaves were enrolled to row warships. First, in 210 BC upon the capture of Cartagena, the Romans sent some captured slaves to the fleet as 'supplementary rowers'[5] although this seems to have been a temporary expedient only, they are not mentioned again. Further, in the naval war between Caesar's heir, Octavian and Pompeius' son, Sextus in 36 BC, Octavian was desperately short of crews for his fleet and thus enrolled several thousands of able-bodied slaves, but they would serve only on condition that he freed them before they boarded the ships, in fact thereby making then Roman citizens.[6]

As an example of the numbers involved, an incident is recorded in 295 BC, during the Third Samnite War, when a levy of some 4,000 Sabellians from coastal Campania and who were classified as *socii navales* (naval allies) on their way to join the fleet, plotted mutiny in support of the Samnite enemy, to whom they were related. The plot was discovered and overcome

but the numbers are impressive.[7] As has been seen, a trireme had a crew of about 200 and this levy was in simple terms, enough to man twenty such ships. Obviously, the Romans would not have had such unreliable men forming the whole crew of a ship and this contingent would have been dispersed to supplement crews from other sources for a greater number and variety of ships than that.

Another example of the allied manpower that was employed, arose in 250 BC, during the First Punic War, after suffering dreadful losses of thousands of men in shipwrecks at sea (in 259, 255 and 253 BC) caused by severe storms; in order to make up the navy's strength, an extra 10,000 men were levied from the maritime allies. This was a considerable imposition in view of the numbers already serving and the fact that the total population of Rome and her allies has been estimated at the time at about three millions.[8] For the Second Punic War, the navy's manpower averaged approximately 60,000 men in total. For the invasion of Africa in 203 BC, the service deployed 160 warships,[9] needing in the region of 30,000 rowers alone and to which must be added the sailors, marines, officers and tradesmen such as ship's carpenters, sailmakers and doctors. One must also add the crews of the fleet of transport ships which carried the army. This was in addition of course to the personnel serving at the same time in the naval squadrons deployed in Spain and Greece.

The furnishing of such huge numbers can only have come from the sources already mentioned, the citizen levy[10] and allied contingents, but virtually every able man must have been needed, bearing in mind that the army most likely had first pick. At the end of the Second Punic War in 202 BC, the navy could be reduced, but even so, retained a hundred ships in commission which needed crews totalling perhaps 30,000 men. Further reductions in manpower took place by the 130s BC, at the end of a succession of wars in the eastern Mediterranean and for which campaigns the Roman navy's ships were augmented by ships from the navies of allied, but still independent Hellenistic states, principally Rhodes and Pergamum.

Roman citizenship to all freeborn men in the Empire, ending the historic differences between legions and *auxilia*, a distinction that, as has been seen, in the close confines of life aboard a ship, had probably been somewhat academic for some considerable time past.

In the traumatic period of disasters that attended most of the rest of the third century AD the navy suffered a lack of recruits and in addition, for the northern and river fleets, local recruitment had to be relied upon more heavily. With fewer men available for transfer, even temporarily from the central fleets, the men and their units became increasingly fixed as garrisons in permanent locations. Pay became increasingly irregular and often stopped altogether in the severe economic climate that beset the empire and which in turn hardly encouraged further recruitment.

From the beginning of the fourth century AD, a system of conscription had to be introduced, together with the re-establishment of regular pay and which seems to have once more provided adequate manpower. There survives no indication of the details of this system for the navy, but it probably followed army practice where each area, town or city was liable to produce a certain number of men each year, in effect a return to the old levy system; in any event, the system and military service itself was unpopular. With the reduction of those fleets, the old system of recruits initially being sent to the Italian fleets no longer prevailed and all new personnel now went direct to their local fleet postings.

The trend of the later empire was towards 'fixing' units including naval units, in various increasingly permanent locations and was allied to the granting of land adjacent to their bases for them to grow food and produce in lieu of regular pay and as a consequence of the vaunted Roman logistical systems which had kept military units supplied becoming less efficient and dependable than before. Whereas in the early Empire men from all over joined the navy and left their homes to do so, thereby breaking their bond with home, which became transferred to the service to which their only loyalty then accrued; with the changes now made, that loyalty was broken, instead the men stayed close to home and having homes, families and land there, felt less affinity for the service as a whole or indeed, to the Roman body politic, upon which they had perforce become less dependent.

In the final decades of the Western Empire therefore, manpower was supplied by conscription and by hereditary enlistment from the sons of navy fathers who pledged them in order to retain 'their' land. The final source of recruits for the navy could have been, as it increasingly was for the army, barbarians from beyond the Empire's borders. From the mid fourth century AD increasing numbers were taken into the army and it seems logical that the navy had its fair share. There survives no evidence as to whether or not such non-Romans served in the navy and if so, in what numbers. In the River and British fleets there may have been some but in the Mediterranean it is less likely, at least prior to AD 419. In that year some Romans, who had been showing the Vandals how to build ships, were executed for an activity which had been made illegal. It would ill serve the empire should barbarians learn how to build ships which could challenge it at sea, as in time they did with disastrous, if not fatal consequences.

WARFARE IN THE earlier ancient world tended to be conducted during the summer campaigning season and ended in the autumn with the onset of poorer weather; also, of course, men had to return to their families, homes and farms to prepare for the winter and indeed, were keen to do so. Men were thus called up for military service in the spring at the start of the season and discharged in the autumn. At sea, for both naval and merchant ships, similar limitations also applied, regardless of whether there was a war or not. The limitations of early ships dictated that from October to March only the rash and foolhardy ventured upon the sea, a practice that continued even with the improvement of ships and navigation, right up to the seventeenth century. It was the Punic Wars of the third century BC, ranging as they did over long periods of time and greater distances and areas than previous wars involving the Romans that changed the age-old pattern of campaigning. The nature of those wars demanded that, although operations were still reduced to the minimum during the winter, armed forces had to be kept in commission throughout the year and men retained to maintain the ships; indeed it is highly likely that a certain number of ships were kept in active commission in winter, even if only as a deterrent. For the Romans at least, this was facilitated by the focus of the First Punic War being on Sicily, so they could always operate from or near to their home ports. To put it into scale however, Carthage is only just over 100 or so miles (160 km) from Sicily, but across open seas

As has been mentioned above, with the period of almost continuous wars of the third and second centuries BC, many of the troops and naval crews were kept under arms continuously, a situation that became formalised into permanent standing forces with the reforms of Gaius Marius between 107 and 102 BC. Men joined the navy and signed up for a fixed period initially likely to have been twenty years, at the end of which they would be discharged, if they did not already have it, with a grant of Roman citizenship. The introduction of permanent service formalised the de facto break from seasons tending the farm and campaigning in each year. The service became their new abode and the regular pay their sustenance, together with any share of plunder that opportunity and an adventurous commander might engineer.

Augustus continued these same basic provisions when forming his Imperial navy, although the possibilities for plunder were much reduced. Rates of pay were kept the same, but the method of funding them improved and they were paid with strict regularity. In order to fund his reformed armed forces, Augustus established a dedicated treasury to underwrite naval finances. Enlistment was increased to a term of twenty-five years, the same as for the *auxilia*. Claudius raised the term to twenty-six years and Marcus Aurelius raised it again, to twenty-eight years. A bounty was paid upon enlistment, one bounty of three gold aurei being recorded in the second century AD.[1] On discharge, many men elected to continue serving beyond their term; others upon discharge, would remain, although not as active crewmen, in ancillary jobs in and around the navy's bases and installations, or would settle in 'navy' towns to run eating and drinking establishments or businesses connected with supplying the service and its personnel. The navy had become their home and family after their long years of service and most of them had become unsuited to a civilian lifestyle far from it, a situation not unknown today.

The rates of pay set by Augustus remained unchanged for approximately a century. The actual rates of pay are not known for the navy but are most likely to have followed the example set by the army. During the first century AD, a legionary was paid 300 denarii per year, out of which deductions were made for his clothing, rations, boots, leather equipment, tent and arms and they would usually contribute to a company funeral club, which took care also of men invalided out of the service, widows and orphans.[2] Despite this, a prudent legionary could save about a third of his salary. This rate can logically be assumed to have also applied to marines, sailors or perhaps leading ratings in the navy who were citizens. As for the non-citizen personnel, they also presumably received similar pay to the army's auxiliaries, namely 100 denarii per annum. Although a lot less, it seems that such pay was still attractive and surviving sailor's wills show that they could amass considerable savings for their eventual retirement.[3] Again, as in the army, discipline was harsh but countered by the guarantee of regular pay, rations and quarters; many sailors could afford to keep a personal slave, some had several. Promotion from the lower ranks brought with it a commensurate increase of pay, calculated as a multiple of the basic rate, the obvious example already seen being the *dupliciarii*, on double pay. A *sesquipliciarius* was another leading rate, receiving one and one half times the basic rate of pay.[4]

Officers' pay rates were obviously higher but almost nothing survives to indicate the relative amounts, a matter that would also help in classifying the exact rank structure. It is only the salaries of the higher ranks that are known, a *navarch*, the equivalent of a senior centurion in the army, receiving some 10,000 denarii per annum, while a fleet prefect and subprefect were rated as *sexagenarii* with 60,000 sesterces per annum, or 15,000 denarii. The praefects of the Italian Fleets were rated as *duocenarii*, receiving 200,000 sesterces, or 50,000 denarii per annum. After his promotion to a rank equivalent to the praetorian praefect, the praefect of the Misene Fleet became a *trecenarius* receiving 300,000 sesterces, or 75,000 denarii per annum. Rates of pay for the lower ranks were increased by Septimius Severus and the men allowed to marry, officially. Previously serving personnel in the army and navy were not allowed to marry, although many men had permanent relationships and even families which could only be legitimised upon the man's discharge from the service. Certificates or diplomas of discharge at the end of service include 'permission' to marry. His son Caracalla gave a pay rise of 50 per cent to all troops and presumably for the navy as well.

In the turmoil of the later third century AD and the economic crises that it brought, the provision of regular pay and rations for the service largely broke down and the men became reliant on payment in kind and their own local efforts in order to maintain themselves and their units. Diocletian stabilised the empire and re-established order and also reintroduced regular pay for the forces, also paying bonuses, for example, on the emperor's birthday. In the reorganisation that took place later, the river fleets on the Rhine and Danube were classified as *ripensis*. Rather than as previously and instead of each operating as one fleet, they were each broken down into individual squadrons. These were linked to and formed part of the frontier garrisons known as *limitanei* at fixed locations and with more limited operating areas. Each unit, with its own land, became more independent and although their sons were conscripted and loyalty secured by the need to defend their homes and livelihood, (thus easing the burden on the Roman exchequer and helping to provide a constant flow of new recruits) they became less flexible as they were increasingly unwilling to operate far from 'home', such relative immobility focusing their loyalties locally.

11 TRAINING

IN THE ABSENCE of any strong evidence we can only make assumptions about the training of the crews of the earliest Roman warships. As has been explained earlier, a core of experienced officers and sailors levied from merchant ships and later, maritime allies, could be augmented by units of soldiers to act as marines and of *capitecensi* to serve as rowers. In the limited number of smaller ships that made up the service in that early period, training of the comparatively small numbers of rowers and rowing crews, is likely to have been 'on the job' as they built up experience under the aegis of rowing masters, the first of whom had learned their craft during the long period of Etruscan domination of Rome.

The limited build-up and use of Roman sea power before the early third century BC with only a few dozen or so ships and probably nothing larger than a trireme, made only modest demands for numbers of crews. This was to change abruptly with the approach of the First Punic War and the huge build-up of the fleet into a world-class one, with the addition, at the start of the war, of twenty triremes and 100 quinqueremes.[1] These ships required a commensurate increase in crews of approximately 33,000 rowers and over 2,300 seamen.[2] This fleet also, of course, needed a corresponding increase in the number of officers, specialist seagoing ratings and marines.

For the marines, training as infantry and with artillery could be accomplished under the aegis of the army's well-developed training regime, leaving them to gain their sea legs once on board.

For the rowing crews, however, a massive new effort was required. It has already been seen that, for an oar-powered ship to work, the rowers must be trained not only to use their oar, but also to row together so that their stroke is synchronised. They were positioned very close together and so had to all move as one, which also maximised the power of their oar stroke as the oar blades all entered the water and pulled together. To achieve this took training and also continued practice.[3]

For the initial training of the rowing crews, 'they placed the men along the rower's benches on dry land, seating them in the same order as if they were on those of an actual vessel and then stationing the *keleustes* (rowing officer, who called the timing), they trained them to swing back their bodies in unison bringing their hands up to them, then to move forwards again, thrusting their hands in front of them and to begin and end these movements at the *keleustes* word of command.'[4] The crews then went aboard the ships for practice at sea before gathering as a fleet, presumably at Ostia, still during the mid third century BC, the main 'home' base of the fleet. Now the fleet had to be exercised in cruising together in formation and in the evolutions necessary for battle, such as forming line ahead and line abreast, as well as changing from one to the other; finally the battle manoeuvres of operating in battle squadrons and the attack formations, *diekplous* and *periplous* had to be perfected.[5]

The fleet then cruised down the coast to Reggio di Calabria, where it paused to refit and embark and install the towers and especially the *corvi* that would give them victory, before venturing to seek out the enemy. That this seemingly cursory training was sufficient is proved by the fleet's performance in the first confrontations against the vastly more experienced, indeed veteran, Punic fleet. Although initially losing some ships, due mainly to poor leadership, when the two fleets met at the battle of Mylae (260 BC) the Romans scored a resounding victory.

The losses of ships and men by the Romans in the First Punic War to storms were horrendous and estimated at nearly 400 ships, with tens of thousands of men and as opposed to only some 200 ships lost to enemy action.[6] Some at least of these losses were due or contributed to by poor seamanship or lack of experience in officers, rather than any want of capability by the crews. As an example, at Camerina, Sicily in 249 BC, the Roman fleet was engaged in combat with the Punic fleet and just as the latter were gaining the upper hand, they suddenly broke off the engagement and ran to the east; their commander had recognised the signs of impending bad weather and gave up the chance of a victory, putting the safety of his ships and crews first. The Punic fleet escaped while the Romans, lacking their enemy's experienced eye, were caught by the storm and their ships wrecked on the nearby rocky shore, only two of their ninety ships and a few transports escaping. By the end of the war, improved training and a regime of continual practice, as well as operational experience, meant that the Roman crews were at the peak of their ability and of the best, the accounts of losses of ships, indeed, of whole fleets, to storm and tempest cease. The fleet that went to western Sicily in 241 BC was maintained by constant exercise, at the peak of performance, which enabled it to win total victory at the battle of the Aegades Islands, which won the First Punic War.[7]

This improved state of affairs was maintained and continued through the Second Punic War and into the second century BC. At the end of that century, however, the navy entered a period of neglect and operated at a very reduced level, although there does not seem to be any diminution in the quality of the crews that continued to function. With the *lex Gabinia* of 67 BC and the war against the pirates, the navy was once more expanded and a regular regime of recruit and crew training re-established, as it had been for the Punic Wars. Although it was now spread over the entire Mediterranean and in several fleets, strong and well-trained naval forces were to be maintained and took part in the several Civil Wars that marked the end of the Roman Republic.

With the establishment of an Imperial navy by Augustus, the service attracted recruits who looked to make a career in it. They received a medical examination on joining and were assigned to a shore-based century, the basic unit of military organisation both for the navy and army. They next received basic training including being taught to swim, some first aid and the use of hand weapons.[8] More specialised training then followed; for example, rowers who had to function in concert were trained initially in rowing frames erected on land as before and before trying their hands in a boat. Ships carried pennant or signal staffs and flew pennants for ranking officers when on board. Signals at sea were made by burnished discs used as heliographs and signal flags and signallers had to be trained to use them with the signal codes.[9] Medical orderlies were sent to the navy's own military hospitals for training and facilities existed for the training of ship's carpenters, boatswains, sailors and all of the myriad other functions needed for the smooth running of the fleet. After training, men were posted to a crew century, which formed the complement of a ship or base. How permanent these postings were is not known, but there must have been circumstances when men moved to other postings, for example, when their ship was retired from service.[10]

As to the marines, their training would continue to be similar to that of the army, their purpose being primarily to act as the assault infantry in boarding or in landing parties. They were trained and practised in the use of the sword, the throwing of javelins and darts and use of the sling, and it seems that they also carried a shortish boarding pike. Archers and artillerymen were also carried but it is not known whether they were additional personnel, or whether the marines and/or the sailors doubled up to perform these functions; the archers are more likely to have been separate units, in the same way that they were in the army. The larger ships that mounted towers on deck could have as many as six archers to a tower and carry up to ten artillery pieces, each requiring a crew of two or three men. In all of this, there is, of course, continuity, possibly refined but fundamentally the same as the training system which had been evolved in the preceding centuries.

12 UNIFORMS AND WEAPONS

THERE ARE A VERY FEW grave stelae which identify those to whom they were dedicated as naval personnel and carry depictions of them. Aside from these there are almost no surviving references or depictions of naval uniforms. Some of the stelae show their subjects in what could be civilian garb, any colouring that may have indicated a uniform having long since worn away. Several reliefs and paintings show marines on the decks of warships, but the convention is almost always followed of indicating a large shield towards the viewer, obscuring the figure but much easier to execute than having to sculpt or paint each figure in detail. A couple of conclusions can however be drawn from these renditions, first, that the marines and the helmsmen wore helmets differing from those of the army, with cheek and face protectors of a reduced type, rather than the almost enveloping varieties of the army helmets. Not surprisingly, in the close confines of a ship and its rigging, no plumes or large crests were worn on the helmets. Second, the marines carried ovoid shields similar in pattern to those of the army but reduced in size to better fit them for shipboard use where some protection would be offered by bulwarks. There are also one or two reliefs that show what seems to be a truncated version of the large rounded legionary body shield, cut off at top and bottom.

In the absence of any other evidence to the contrary, one can only assume that the patterns of naval uniforms followed those of the army, with adaptations where necessary. Thus the army's well-known ventilated boot or *caligae*, normally studded with hob-nails and which would provide little or no grip on the deck of a ship, to which they would also inflict severe damage, were likely to have been replaced by a version with perhaps a rope sole. The standard tunic issued to rowers would have had a reinforced seat, for obvious reasons. Experience in the operation of the replica trireme *Olympias* showed that the rower's cushions alone were insufficient and the reinforcing of the seats of the rower's clothes followed.[1]

Throughout the navy's history, the basic uniform for all hands would start with that universal garment of the ancient world, the tunic, worn to the knees and with sleeves to half-way down the upper arm and worn with a belt or *cingulum*. There are many depictions of sailors on merchant ships so dressed, in addition to the men shown on ships on monuments such as Trajan's Column in Rome. Boots or shoes and a cloak for use ashore, along with underwear completes the basic kit. The tunic was of wool and of finer quality for the officers. The garments are likely to have been left in their natural off-white for seagoing use to obviate the bleaching effect of salt water on dyed cloth. Perhaps a coloured tunic was issued for parade or shore use. For the most part of the navy's heyday, between approximately 300 BC and AD 300, the climate was warmer than today and as the sailing season was limited to the hotter summer months, this simple outfit sufficed for seamen and rowers, at least for the Mediterranean. For winter wear by about the first century BC, breeches reaching to mid-calf were introduced, probably with a long-sleeved tunic, but it was not until the late third century AD or early fourth century AD, along with the colder climate, that trousers and long-sleeved tunics came into more general use and issue, especially in the northern theatres of operations.

Regarding uniform colours, it is unknown whether any marking or pattern was included to signify a particular ship's crew or a particular fleet or other formation such as a century or squadron. There is one definite account of naval uniform colour from very late in the empire, related by Vegetius in his military

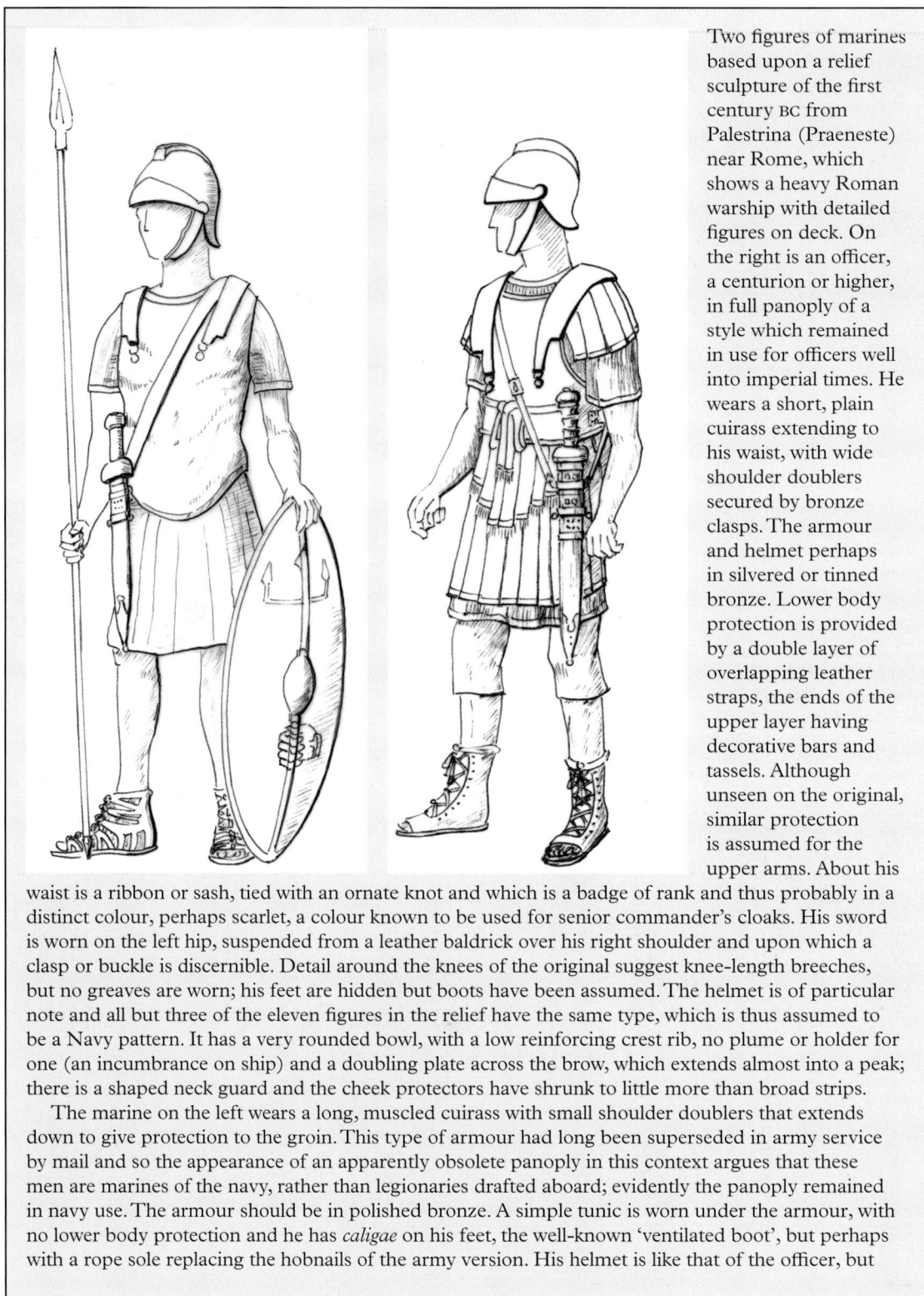

Two figures of marines based upon a relief sculpture of the first century BC from Palestrina (Praeneste) near Rome, which shows a heavy Roman warship with detailed figures on deck. On the right is an officer, a centurion or higher, in full panoply of a style which remained in use for officers well into imperial times. He wears a short, plain cuirass extending to his waist, with wide shoulder doublers secured by bronze clasps. The armour and helmet perhaps in silvered or tinned bronze. Lower body protection is provided by a double layer of overlapping leather straps, the ends of the upper layer having decorative bars and tassels. Although unseen on the original, similar protection is assumed for the upper arms. About his waist is a ribbon or sash, tied with an ornate knot and which is a badge of rank and thus probably in a distinct colour, perhaps scarlet, a colour known to be used for senior commander's cloaks. His sword is worn on the left hip, suspended from a leather baldrick over his right shoulder and upon which a clasp or buckle is discernible. Detail around the knees of the original suggest knee-length breeches, but no greaves are worn; his feet are hidden but boots have been assumed. The helmet is of particular note and all but three of the eleven figures in the relief have the same type, which is thus assumed to be a Navy pattern. It has a very rounded bowl, with a low reinforcing crest rib, no plume or holder for one (an incumbrance on ship) and a doubling plate across the brow, which extends almost into a peak; there is a shaped neck guard and the cheek protectors have shrunk to little more than broad strips.

The marine on the left wears a long, muscled cuirass with small shoulder doublers that extends down to give protection to the groin. This type of armour had long been superseded in army service by mail and so the appearance of an apparently obsolete panoply in this context argues that these men are marines of the navy, rather than legionaries drafted aboard; evidently the panoply remained in navy use. The armour should be in polished bronze. A simple tunic is worn under the armour, with no lower body protection and he has *caligae* on his feet, the well-known 'ventilated boot', but perhaps with a rope sole replacing the hobnails of the army version. His helmet is like that of the officer, but in polished bronze. He is armed with a short, boarding pike and a dished, ovoid shield, held by a single, central handgrip and being a smaller version of the army's body shield. Curiously, no sword is shown; as every Roman fighting man carried one, this must be an omission, or perhaps a detail that has been lost over time. From the sarcophagus of a Roman officer of similar date, soldiers in similar garb are shown with swords, suspended by baldricks in the same way as that of the officer, but on the right side.

The *praefectus classis*, right, was the commanding admiral of an imperial fleet. Prefects of the two Italian (home) fleets ranked with the Praetorian Prefect and their uniforms and panoply would be of the very highest quality. The illustration shows a fleet prefect of the first century AD and the armour is based upon a statue of the emperor Vespasian from the Temple of the Augustales at Misenum, which is assumed to be showing the emperor in his capacity of admiral. The panoply consists of a full, muscled cuirass extending to the upper groin, with a head of Medusa in raised detail on the chest. Small shoulder doublers are added, together with the ornately tied ribbon denoting rank, perhaps in purple. The lower body is protected by a single layer of leather straps ending in tassels, white or red with gold ends perhaps and with corresponding straps covering the upper arms. The cloak is gathered and held on the left shoulder, as commonly shown in statuary and probably in scarlet, the mark of a senior commander. The tunic would be of the finest weave and ornate, and laced open-toe boots complete the ensemble. The finely decorated sword is slung from a baldrick with a richly furnished scabbard and of a type with an eagle-head pommel exclusively for generals and upward, known as a *parazonium*. The helmet is from a sarcophagus of the first century BC of a senior Roman officer.

Left, an officer of the *Classis Pannonica* in 'walking-out' clothes, a tunic with long, loose sleeves, knee-breeches and a cloak draped about his left shoulder and secured on his right by a clasp. He may be wearing armour: there are very faint markings over the right upper arm and the navel indentation could indicate a 'muscled' cuirass with shoulder protectors. He wears shoes (*calcei*). His thick military belt is secured by a ring-type buckle and the thickness may indicate that it was covered by decorative plates. There is a curious apron-like flap on his left thigh, which is unclear. From an altar dedicated to Neptune from Vindobona, dated to AD 279. Romer Museum, Vienna. (*Author's photograph*)

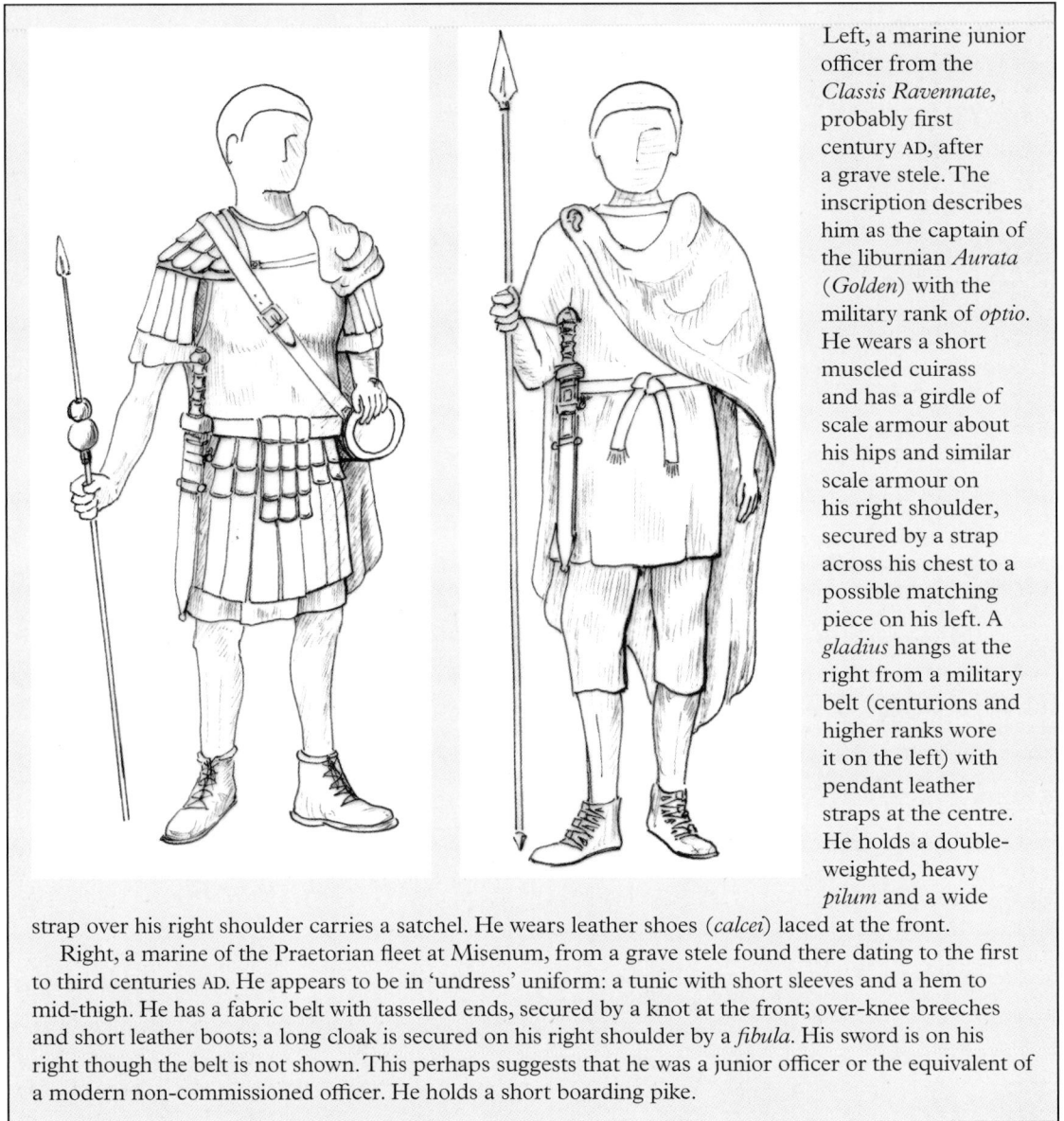

Left, a marine junior officer from the *Classis Ravennate*, probably first century AD, after a grave stele. The inscription describes him as the captain of the liburnian *Aurata* (*Golden*) with the military rank of *optio*. He wears a short muscled cuirass and has a girdle of scale armour about his hips and similar scale armour on his right shoulder, secured by a strap across his chest to a possible matching piece on his left. A *gladius* hangs at the right from a military belt (centurions and higher ranks wore it on the left) with pendant leather straps at the centre. He holds a double-weighted, heavy *pilum* and a wide strap over his right shoulder carries a satchel. He wears leather shoes (*calcei*) laced at the front.

Right, a marine of the Praetorian fleet at Misenum, from a grave stele found there dating to the first to third centuries AD. He appears to be in 'undress' uniform: a tunic with short sleeves and a hem to mid-thigh. He has a fabric belt with tasselled ends, secured by a knot at the front; over-knee breeches and short leather boots; a long cloak is secured on his right shoulder by a *fibula*. His sword is on his right though the belt is not shown. This perhaps suggests that he was a junior officer or the equivalent of a modern non-commissioned officer. He holds a short boarding pike.

treatise dating from the early fifth century AD.[2] In it he refers to scouting ships in use by the northern fleets operating from Britain and the northern coasts of Gaul. These ships were used for scouting, interdiction into enemy territory and similar clandestine activities for which some form of camouflage would have been an asset. Accordingly, Vegetius relates, these ships were painted sea-blue, with the sails and even the rigging dyed to match and to complete the scheme, the crews, seamen and marines alike, wore similarly coloured uniforms. Whether similar measures were adopted in other forward areas such as the Lower Danube, Vegetius does not mention, but one can assume that, the principle being known, camouflage would have been used wherever of use.

Without their distinguishing helmet crests and plumes, officers must have had some other system of indicating rank. For very senior officers

A marine in 'light order', second century AD, after a relief of this period showing marines on a ship deck in the act of throwing a volley of javelins or *pilae*. He is in a tunic, but no body armour and wears his military belt, with decorative metal plates. His sword, on his right, is hung from a baldrick, just visible on the relief. His helmet is of a type similar to a grave stele of a Roman proconsul of Bithynia (first century BC) and a similar type on the Arch of Constantine in Rome (fourth century AD). He wears *caligae* with rope soles and carries a reduced-in-size version of the army's rectangular body shield, with a central metal boss. He is in the act of throwing a weighted javelin, a weapon with a range of 30 to 50 yards (27–45 m), which could be increased to about 70 yards (64 m) by the use of an *amentum*, a long thong with a thumb-loop on one end and a leather cup on the other into which the end of the javelin was placed; the thong was wound around the shaft and when thrown, increased the throwing length and imparted a spin.

of senatorial rank, under both the Republic and the Empire, a broad purple stripe on their uniforms, as well no doubt as very smart armour, would be worn. In Republican times, the Senate, when appointing a supreme commander of a large force, would present him with a scarlet cloak as a badge of office.[3] There is no reason to suppose that this tradition did not continue and, for example, be worn by a fleet prefect. There is a reference to the granting of such a scarlet cloak by the Emperor Commodus to Clodius Albinus (emperor AD 192–193) in the late second century AD.[4] One badge of rank that is attested for the navy (as well as being in use by the army) from the grave stele of a marine, is a ribbon tied above the waist and knotted at the front with the ends which ended in tassels either hanging or looped and tucked into it; different knots were used, perhaps according to rank; colours are not known but perhaps they also varied according to rank. Something more noticeable, such as a coloured band painted around the helmet (as used by the Greeks for decoration) could have been a simple answer to the problem but it must be emphasised that this is pure conjecture and is not attested.

As for armour, once again the navy most likely followed the fashions of the army. In a second or first century BC frieze of marines on deck shown in the act of hurling javelins or *pilae*, they are equipped with helmet and shield, but no armour;[5] conversely, in a relief from Palestrina, the marines are shown wearing 'muscled' breast and back plates, full armour but without greaves. Both reliefs date from approximately the same period (first century BC) and whereas the former perhaps depicts the marines in 'light order', the latter shows them heavily equipped but in armour of a type that by that time had been superseded in general use in the army by mail armour. Vegetius for his part strongly recommends the use of armour by marines.[6]

Curiously perhaps, none of the men shown manning the many ships depicted on the columns of either Trajan or Marcus Aurelius in Rome are wearing armour or even helmets, but are shown only in tunics, unlike those shown ashore, who are.

Before the foundation of Rome, traditionally in 753 BC, various forms of body armour were known and in use in Italy, the bronze cuirass and also scale armour, made up of many small and overlapping metal plates fixed to a backing garment; there was also a home-grown, simple panoply consisting of round or rectangular bronze chest and back plates held in place by leather straps, the latter cheap to produce and widely used. The Etruscans evolved a cuirass made up of multiple bronze plates fixed to a backing of perhaps linen or leather and very similar to the Greek linen cuirass. The simple panoply was developed and expanded into the 'triple-disc' type of matching breast and back plates, linked by broad bronze bands over the shoulders and at the sides and worn with a broad bronze military belt. A variation used rectangular 'muscled' breast and back plates. Judging from the many examples of this type of panoply in museums in southern Italy and dating from the Punic Wars period, these types were popular among Roman troops (there is even a fine example in the Bardo Museum, Tunis) all of which would suggest that they were in use by shipboard troops; these panoplies are not cumbersome yet provide a moderate degree of protection and seem ideally suited to shipboard use.

The full cuirass in bronze was expensive to produce and its use therefore restricted to officers and some heavy infantry. Mail, developed on the Celtic fringes of northern Europe, does not appear to have been in use by the Romans until the Second Punic War and but was in widespread use until the late first century BC. It then remained in use until and beyond, the end of the Empire. Scale armour, apparently more popular in the East, became more common in Roman service with the induction into their forces of troops from the eastern Mediterranean area from the early Empire onward. Segmented plate armour was in general Army use, alongside the other forms, from the mid to late first century AD to the end of the third century AD, possibly later; its use by the navy is not yet attested however.

Finally and more problematical, due to the lack of surviving examples, the extent to which armour of hardened leather was used remains a mystery. Examples in the material, of chamfrons to protect horse's faces have been found dating to the third century AD and more pertinently, a leg guard has been recovered from a shipwreck (date unknown). In all of these cases the use of the various types of armour by the navy can only be assumed, the Palestrina relief referred to above and an early imperial grave stele from Ravenna, being the only definite portrayals, so far found which, show marines in armour.

The full range of personal weapons used by the army were also available for the navy and swords, spears, javelins or *pilae* are illustrated in reliefs. There also appears to have been a shorter boarding pike for marines and the use of grapnels, axes, slings and archery is attested. Vegetius[7] also mentions the use of the *fustubulus* or staff sling and lead weighted darts, the latter an introduction of the late empire. There was also the *amentum*, used to increase the range and accuracy of a javelin, consisting of a leather thong, about three feet (915 mm.) long, with a thumb loop on one end and a cup for the butt end of the javelin on the other; the thong was wound around the shaft to impart a spin when thrown.[8] A specialised naval hand weapon was developed for Caesar's campaign against the Celtic Venetii of Brittany in 56 BC and was a sickle-like blade mounted on a long pole.[9] It was used to snag and cut the rigging of enemy sailing ships to render them helpless, it continued in use at least until the late fourth century AD.[10]

Personal weapons evolved and changed during the long period under consideration, with for example, the early Etruscan and Greek style swords being succeeded by the famous *gladius hispaniensis*, which in turn changed form from the earlier, very fine-pointed types to the broader, short-pointed form in imperial times; this was in turn superseded by the *spatha* or longer sword, originally issued to cavalry only. Such changes are well documented and attested for the army but, can only be posited for the navy. With the increased concentration of manufacture into dedicated arms and armour factories from the late third century AD[11] standardisation of types for the Roman forces presumably increased and it follows that issue of such equipment to all units included naval units.

PART V   SERVICE LIFE

13   FOOD AND DRINK

THE ESSENTIAL FUEL for the human engine upon which ancient warships depended, food and drink was a matter that needed careful planning and arrangements. Provisioning a naval force, especially one which became as large and widespread as that of the Romans, always presented problems. These were exacerbated by the operational areas, types of ship in use at the time and the methods and speed of transport, as well as the productive capacity of ancient agriculture. The matter can be conveniently considered in two parts.

Ashore

Up to the end of the First Punic War in 241 BC, Roman naval formations operated from their Italian homeland and around Sicily. They operated from, or had ready recourse to, home ports and always within easy reach of the army. The question of supply was therefore the same as that for the army, the same organisation and furnishing of victuals supplying both simultaneously and over relatively short, internal lines of supply. Sicily itself was a major producer of wheat and other produce and so long as the Romans controlled most if it, which they did for most of the war, it could provide the major part of the rations required by the Roman forces there. After their initial naval victories, the Romans could also control and secure the short sea routes of the Tyrrhenian Sea, between Sardinia (another major producer), Sicily and the mainland. This possession of the two islands during the Second Punic War (from 218 BC) enabled the Romans to feed themselves and their allies despite the loss of agricultural production on the Italian mainland caused by the sixteen years of the activities of Hannibal and his army there.[1]

From the Second Punic however, naval formations had to operate far from home, notably in Spain and Greece and eventually in Africa. In these operations they were not always operating with the army, particularly in Greece where there was no Roman army. The fleet had therefore, to develop its own supply systems for these detached units. For the Spanish theatre, this was not difficult, as one of the principal duties was the escorting of supply convoys between bases in Italy and the Roman forces in Spain. Ships could thus victual at home ports in Italy and use part of the supply convoys to maintain adequate stores at their Spanish destinations, to cover operations in that theatre. Bases at harbours such as Tarragona, Sagunto and, after it was taken, Cartagena, secured termini of Roman lines of supply, both for receiving supplies from home and to support naval operations along the littoral. A similar situation existed after the invasion of Africa in 204 BC, where stores were brought by the convoys and could be maintained alongside those of the invading army.

For the Greek theatre however, local victualling had to rely on local allies once the fleet had left the Adriatic and Ionian areas of Epirus and Corfu which were Roman occupied and controlled. In the Aegean the Romans seized and secured the island of Aegina, near to Athens, and used it as their base. They were not acting in isolation and as well as their Aetolian allies on land, they had the support of the fleet of King Attalus of Pergamum and were able to rely on the supplies emanating from there.

With the establishment of the Imperial fleets, the situation coalesced into a status quo that would more or less endure for the next several centuries. Permanent fleets with permanent, established bases and their own treasury meant that there were similarly permanent arsenals. Excavation at naval bases has revealed stores, cisterns, granaries and such facilities for the maintenance and victualling of those fleets. There are examples at Misenum, Isaccea (near the mouth of the Danube) and Xanten on the

Barrels. The amphora was the universal container of the ancient world but, the wooden barrel could carry a similar quantity in a container that weighed very much less, a valid consideration for warships. Its use spread from northern Europe where it originated and is here seen, left, aboard a lighter on the Danube. Trajan's column, early second century AD. Below: barrels are used to convey cargo on a river lighter also of the second century AD, being towed on the River Durance in southern France. Avignon Archaeological Museum. (*Author's photographs*)

Rhine. Such bases held stores of wheat, hard cheeses, dried fruit, cured meats and salt fish.[2] Bakeries produced biscuit as well as bread and groats; amphorae were filled with olive oil, olives, honey, preserved plums, figs and dates, lentils, dried beans and rice.[3] Fresh fish, meat, fruit and vegetables could be purchased locally. Later, from the early fourth century AD, static border units were granted plots of land adjacent to their bases, upon which to grow their own food, the old system for the supply of which had become severely disrupted. As the border fleets evolved into a number of smaller formations, subject to more local command, they also had increasingly to rely on their own efforts for provisions. This can also be seen to apply to the *Classis Britannica*, whose bases also each had their own agricultural holdings.

Afloat

The very design of the warships of the ancients militated against the carrying of any but the bare minimum of stores, having little or no free space after cramming the maximum number of rowers in the minimum space. The basic principal from the earliest of times was that a ship had to put into shore at night, to disembark the crew so that they could prepare and eat their food and rest. There is not known to have been provision on board the ships to have done these things, any eating or resting while afloat or under way, having been strictly ad hoc. Once room had been found for the ship's gear, spars, sails, cordage, spare oars, weapons and such like, the next priority was for the stowing of drinking water. The majority of the crew were rowers, and in all but the smallest ships, they were working below deck in a box with limited ventilation, mostly in the hotter months of the year. Hard physical work in a hot, confined space caused them to sweat profusely and a constant supply of drinking water was an absolute necessity. It was found in the operation of the replica trireme *Olympias*, that an allowance of one litre of water per rower per hour was required.[4] Another study estimated a bare minimum of 4 pints (2.5 litres) of water per man per day.[5] Even allowing this minimum amount, then the 170 rowers of a trireme will need 85 gallons per day; with one gallon weighing 11 pounds, a ton will last a maximum of only two and a half days. The 250 rowers of a quinquereme will consume over a half a ton of water per day.

Water was carried in large terracotta containers (*dolia* or *pithoi*) and in animal skins, to be decanted into smaller vessels for distribution to the rowers as they worked. From the second century AD, if not earlier, the wooden barrel, an invention of northern Europe, became more widespread in use, being lighter for the same carrying capacity than the equivalent terracotta container.[6] Whatever type of container was used, it had to be accessible while the ship was under way without disturbing the rowers and so that their contents could be distributed. In the limited space available and to keep weight down, only a limited supply could be carried and needed to be replenished frequently. The bigger ships were in no better position as, although larger, they had more rowers and as restricted a storage space pro rata.

Food for the crew was a similar problem. Earlier Greek, and presumably other, cruising ships would be beached at night and the crew sent off to buy (or steal if they were buccaneering) provisions from the nearest settlement. Except in the larger and friendly ports, this was not an option for Roman warships, let alone for squadrons of them. There is a reference to 'oarsmen with pay and rations for thirty days'[7] from which it can be seen that the amount of a day's pay and rations were defined; it does not say, however, how many day's rations were to be embarked aboard ship. On another occasion however, ships were to be stocked 'with ten

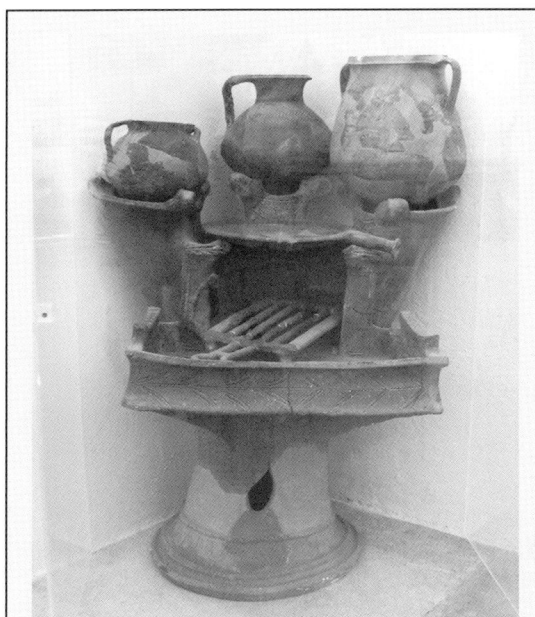

An ancient cooking range or *thermospodium*, made from terracotta but other, smaller examples have been found made from metal. Such devices could be carried aboard a ship and taken ashore to cook food for the crew. Archaeological Museum, Delos. (*Author's photograph*)

days' cooked food.'[8] Again, reference is made to 'basins and hand mills to equip forty warships'[9] from which it can be deduced that ships carried equipment to enable food to be prepared and that grain was carried, to be ground into flour; it does not say however, that this was a shipboard activity. Indeed there is a further reference to 'troops and marines messing together' in Britain, that is ashore.[10]

For the invasion of Africa in 204 BC, which was, of course, to be an extended campaign, the stores loaded for the invasion fleet included food for forty-five days, to include cooked rations for fifteen days and the same number of days' supply of water for men and animals as food.[11] The crossing to Africa in fact took just over twenty-four hours and the supplies must therefore have been destined mostly for the invasion army and for that matter, to have been carried in the transport ships. Nevertheless, no distinction is made between warships and transports in the orders and although probably including the close escorts, there were other, covering squadrons that need not be so burdened. The crossing had been quick, an earlier raid on the North African coast had taken three days,[12] whereas another such cruise had taken thirteen days.[13]

Ships must normally have carried food and drink for their crews, other than the mandatory water supply, of a limited nature, perhaps three to five days' worth as a matter of course,[14] or such amount as suited the voyage or patrol being undertaken as above, bearing in mind always the limited space and the number of suitable intermediate stops available. Cicero, when travelling in 51 BC to take up his appointment as governor of Cilicia in Asia Minor, was sent by small warship, which took two weeks for the journey, stopping each evening at a suitable port, for the night, the ship therefore needing only one day's rations at any time.[15]

Rations

What foods might seagoing rations include? As the ships, or at least some of them, were equipped with hand mills, they must have carried rations of grain, which could be ground into flour to make bread or porridge (*pulmentum*) once ashore. Such porridge or pottage had been a Roman staple since the earliest of times and was basically very simple to prepare, simply add ground grain to water and heat; to this could be added meats or vegetables, oil or wine, there were many recipes, most of which would be suitable for the crew of a ship to be able to prepare, being a basically simple fare but which also required water which had to be carried or obtained.[16] As a guide to amounts, an infantryman of the third century BC, received a ration of a half-bushel of wheat per month (28 lb, 12.7 kg),[17] this would produce approximately 20 lb of flour, (9.1 kg), enough for 26 lb (11.8 kg) of leavened bread. A military speciality was bread rolls baked on a spit.[18] There were alternatives as food staples, notably groats, where grain was first roasted on a heated threshing floor or large hearth stones, divested of the chaff, then pounded into a coarse flour (or groats). This flour was now digestible and if mixed with a little water, formed a stiff paste. The same process can be applied to wheat, barley, lentils, chickpeas or beans and of course, the results mixed to form a more nutritious result. Such grain pastes were known from neolithic times and still have modern equivalents (such as polenta).[19] The paste (*puls*) keeps well and Pliny suggests placing it in a container sealed with flour and bran, (ideal for shipboard use) or mixing it with olive oil or honey and seasoning.[20] The paste could be diluted to make a porridge, eaten as is, or placed on a griddle to produce a wholesome, unleavened bread. Wheat flour was also mixed into a dough with water and dried in strips, but required boiling water to become edible (pasta). There is a reference to Greek rowers mixing barley meal with oil and wine, which they ate while rowing.[21]

Another staple food of the seaman was ship's biscuit, a staple food for mariners for centuries. Bread, baked twice (*bis coctus*, hence biscuit) to remove moisture and kill bacteria, is a foodstuff with a very long life if kept reasonably dry. The ancients were unaware that they were killing bacteria, but the long-lasting food was useful, though preservation was not such a serious problem for the comparatively short voyages undertaken: the longest regular journey was that of the grain carriers from Alexandria in Egypt to Italy, a voyage of, on average, between nine days and two weeks.[22]

Whether wine was carried aboard as part of ship's rations is not known for certain. The first priority, as has been seen, was to carry water but wine was such a fundamental part of the Mediterranean diet that not to have carried at least a small supply is inconceivable and indeed proven by the account of Thucydides.[23] The ancient Greeks and Romans usually diluted their wine with water, so a little could go a long way, especially as the wine could be of variable quality, everything from excellent to a mild vinegar; the latter in fact, makes a refreshing drink when diluted.

ANCIENT WARSHIPS, as has been seen, made provision only for movement and fighting; there was no accommodation, all the space in a hull being designed around and used for those two functions, As such, they could not and were not intended to remain at sea for extended periods. While at sea, the sailing rig would be employed whenever the wind was favourable to augment the rowers but it was they who always provided the prime motive power. In multi-reme ships the remes could be rested in turn while the remainder rowed. Below decks the heat and sweat generated by hundreds of toiling rowers, especially in the hot Mediterranean sun required that provision be made for ventilation and a flow of fresh air to enable the men to function. This was provided by large, open deck hatches, which also allowed entry and exit and which could be closed by hatch covers when the ship was in action. Canvas wind scoops could be rigged over the hatches to give a greater airflow to below deck. Again, as has been seen, many of the surviving ship illustrations show an open row or tier between the deck and the topmost row of oars with what could be an open lattice framework which is interpreted as a ventilation course, which again could be closed by covers for combat; finally, the undersides of oarboxes could be left open to assist air flow.

One of the first jobs for a crew when reaching a harbour was to swab down the inside of the hull to wash out the accumulated sweat and detritus of a voyage and to keep the ship sweet. In the *Olympias*, this was found to be desirable every four days.[1] There remains the question of the provision of drinking water for the rowers at their benches. The installation of water containers in a ship has been considered but there is no record of a rank of 'water boy' or such like. The duty of taking drinking water to the men at their oars was most likely fulfilled by including within the crew numbers a few more rowers than there

were oars, these supernumeraries or reserve rowers[2] being available therefore to dispense water and also to take over if a rower needed to visit the heads or to replace any man who was taken ill or injured. Once again, this is a suggestion only and although logical and fitting within the known structure, is not attested.

There is no record of the sanitary arrangements on board these ships, for the crews or their passengers. In many illustrations of both war and merchant ships a stern platform is shown projecting beyond the rear of the ship hull, overhanging the wake and it is reasonable to surmise that they acted as the 'heads' for the ships, certainly later, medieval ships had projections built outboard of the stern for this purpose.

Transporting troops

There were many occasions when warships were used to transport troops, for example from Sicily to North Africa and across the Adriatic. In addition to the hundreds of men forming the crews, decks would have been crowded with troops and their equipment. It is not easy to imagine what it must have been like for these troops on the ships, especially for extended voyages. A ship such as a quinquereme, with an upper deck extending to approximately 130 feet by twenty feet (39 by 6 m), reduced by being tapered at the fore and aft ends, was further reduced by having to deduct space at the stern for the officers and helmsmen, in the bow by space for the crew to work the anchors, cables and rig and amidships to give room for the mast and rigging and for the seamen to work them; more if an *artemon* or foremast was rigged, which they usually were.[3] Additionally there were catapults and towers on the larger ships and the many large deck hatches left open, all of which reduced the deck space available for the troops and their kit by at least a third, leaving perhaps 1,300 square feet (123 sq. m) to accommodate them. Each man had his helmet, armour, shield, weapons and personal pack of cloak, food, water bottle, mess tin, etc., on this moving, probably wet and comparatively tiny space for hours on end. In the medieval period it was the practice that 2,000 square feet (189 sq. m. of deck space could house 250 men, or a space eighteen inches by five feet six inches (45 cm. x 1.8 m) per man to lie down. In light weather, awnings could be rigged to give some shade and Caesar mentions leather awnings for rougher conditions.[4]

That such numbers were transported is reported in the ancient sources, for example, 4,000 men were conveyed on fifteen ships for a shortish journey between Greek ports, an average of 266 men per ship;[5] on another occasion, thirty ships were used to carry 7,000 men, or an average of 233 per ship;[6] finally, thirty quinqueremes disembarked the whole First Legion of perhaps 4,500 men, a comparatively spacious average of 160 per ship.[7] There survives an account of what was considered to be a rapid voyage by a Roman squadron transporting marines from Brindisi to Corfu, a distance of 110 miles (176 km) in eleven hours, running under sail and with the decks crammed with troops. It must have been not unlike holding your luggage on a flight from London to Los Angeles, with as little or less room and having to stand for a lot of the way. This voyage was fast enough to warrant mention and other, much slower passages are recorded with rates of travel varying from only two and a half knots to a comparatively rapid average of four and a half knots.

Noise

Below decks hundreds of oarsmen laboured for hours on end, obviously pacing themselves for endurance, but nevertheless, engaged in hard, mindless, physical work. The sheer noise, even with the men quiet, of a 150 or so oars being worked together, each straining against its thole pin and retaining strap, allied to the creaking and noise of movement of the hull of the ship, together with the sound of the water,

the shouted and trumpeted orders and the monotonous beat of the *portisculus*' hammer marking the timing of the oarstroke, must have been numbing. Assuming that the normal station for the *pausarius* (rowing master) was at the rear of the aft-most or stroke oarsmen, facing them and just below the quarter deck, where he could receive orders from the captain, he then had to give his own orders to the rowers. A shouted order could be repeated, transmitted by senior rower ratings posted at intervals along the remes, each to his own squad. Alternatively, orders could be transmitted by a musician; the ranks of *bucinator* (trumpet player) and *cornicen* (horn player) are attested for the navy from grave stelae. Both were bronze wind instruments with distinctively different sounds and as for the army, a combination of instruments and calls could convey a large variety of orders with sounds that would carry above the noise below decks. For any other than experienced hands, the repetitive and regular 'pulse' of the whole ship as all of those oars dug into the water at the catch, at the same time must have added to the general discomfort, especially for those unfortunate enough to have been poor sailors.

Manning the ships

with crews numbering in the hundreds, embarking and disembarking them, must have been processes requiring considerable organisation and discipline, together with a great deal of practice. Trials with the trireme *Olympias* showed that the ship's 170 rowers could board and run out their oars in as little as one and a half minutes and retract the oars and disembark in the same time. In an emergency simulation, the crew managed to get out and over the side in only twenty-four seconds.[8] For a fully decked ship in action, with protective covers over the hatches and ventilation course and the greater numbers in say, a quinquereme then, even allowing ten times as long, would still mean that the crew could evacuate in only four minutes; it is difficult to conceive even the most severely damaged ship taking less than that to settle enough to trap the men below. That this was generally so can only be demonstrated by reference to contemporary battle reports, where such instances where men were trapped below were rare enough to merit especial mention.[9] As a rule therefore, it can be assumed that men were able to escape a ship stricken in battle and if they could not capture an enemy vessel or swim to one of their own, they had to seek salvation by clinging to any sufficient piece of floating wreckage.

The position was very different when it came to ships assailed by storms. Although most larger warships carried a small ship's boat, there is no evidence that any form of lifesaving equipment was available, other than large pieces of cork, which it seems, were commonly carried for use as, for example, markers for anchor cables. Lucian relates an occasion when they were thrown to men who had fallen overboard.[10] Apart from this, or being fortunate enough to be washed on to a friendly shore, they drowned. The death toll from storms could be enormous; in the First Punic War the Romans lost some 700 ships, of which some 400 succumbed to storms;[11] in the worst incident, in 255 BC, the main Roman fleet, laden with booty and almost as many captured Carthaginian ships, with prisoners and prize crews aboard, was caught between Malta and southern Sicily by a violent storm and from a total of nearly 400 ships, only eighty of the warships and a few of the transports managed to limp into harbours in Sicily. It was

and remains the greatest known loss of human life in a single incident of shipwreck in history, as many as 100,000 men being lost beneath the waves. It was not only the Romans who suffered, in 205 BC, a Carthaginian supply fleet, with reinforcements and supplies for Hannibal, was blown all the way to Sardinia, where the survivors were promptly Captured by the Romans there.

As to moving the numbers involved in manning a whole fleet, of Octavian's fleet preparing for the battle of Actium, it was reported that 'the men went aboard and formed up for battle'. This somewhat bland statement belies the scale of the operation, approximately 200 ships with rowers, marines, sailors and extra troops drafted aboard for the battle, involved between sixty and seventy thousand men, the population level of a fair-sized town. Embarkation in fact started before dawn, nevertheless, to get the ships to sea and in position and formed for battle, a distance of less than two miles, took until midday and that was with experienced crews.[12] For Scipio's invasion of Africa, the ship crews were ordered to man their ships and stay aboard while the army and it's supplies and transports were loaded , for an army of (the sources vary) between 12,000 and as many as 30,000 men, this must have taken several days.[13]

15 RELIGION AND SUPERSTITION

PERHAPS IT IS the sheer awesomeness of the sea that has always inspired sailors to be superstitious and Roman sailors were no exception. One of the earliest and most universal manifestations of this is the *oculus* (an eye, painted or fastened to the prow of a boat or ship to help her to 'see' her way). It was considered bad luck to sail on a Friday and Roman ships adopted a patron deity as their guardian and after whom the ships was commonly named.[1] A shrine or statuette to the deity was placed in the stern of the ship and on boarding, a crewman would turn towards the stern and offer a prayer or salutation for a safe voyage; a custom that survives in today's navies, of saluting the ensign at the stern upon coming aboard.

Divination

The art of divination, foretelling the future by observing natural phenomena such as eclipses and the flight of birds as well as, particularly, from examining the entrails (especially the liver as the biggest of the organs) of sacrificial animals, had come from the Etruscans.[2] From those people's long and close association with Rome, the practice had become deeply embedded in the Roman way of life, where it was practised by a soothsayer known as a *haruspex* or *augur*; as late as the mid-first century AD, the emperor Claudius established a college of augurs. The omens were consulted before any major military undertaking a process that included the committing of fleets. Most famous was the case of the consul Publius Claudius Pulcher, commander of the fleet in Sicily in 249 BC, who planned an attack on the Punic fleet at Trapani. When the augurs told him that the omens were bad as the sacred chickens would not eat, he replied that 'if they will not eat, let them drink' and promptly had them thrown overboard; the venture was a disaster and the Roman fleet was destroyed by the enemy in

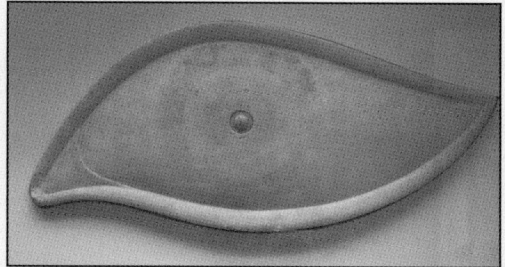

One of the earliest of devices to decorate a ship and one that is still in use is the *oculus* to help a ship to 'see' its way home. This example is in fine marble for attachment to the prow of, it is thought, a trireme. Fifth century BC. Piraeus Archaeological Museum. (*Author's photograph*)

the ensuing battle, the only one of the war that they lost.[3] A more positive result accrued to the future emperor Titus who was assured by the priest after 'reading' entrails of a clear passage and a calm sea.[4]

Gods and goddesses

Castor and Pollux (the Dioscuri) were the sons of Zeus and Leda, wife of Tyndareus.[5] They joined Jason and the Argonauts in their quest for the Golden Fleece. When their ship, the *Argo*, was caught in a storm, two flames sent by Zeus descended from the sky and hovered over their heads: this was believed to be the origin of St Elmo's Fire.[6] The twins thus became established in Greece as patrons of sailors and were then adopted by the Etruscans. They were claimed to have appeared at the head of Roman cavalry, which was victorious and thus made the progression into the Roman pantheon, but as horsemen in silver armour. Their connection with the sea and sailors continued among the Romans, who regarded the lightning (St Elmo's Fire) playing in the rigging during electrical storms as a sighting of the Dioscuri themselves.

Above left, the Dioscuri, Castor and Pollux, the divine horsemen, whom sailors believed appeared as St Elmo's fire when lightning played in the rigging of their ships. Second century AD. Metropolitan Museum of Art, New York.

Above right, he cult of Mithras, with its strong symbolism and the camaraderie encouraged among its followers, appealed to and became widely followed among the military. Head of Mithras, early third century AD. Archaeological Museum, Arles.

Right, a small altar dedicated to the sea god Neptune by a centurion C. Vibius Celer. Late first–early second century AD, from the Danube frontier, near to Vienna (Vindobona). In view of the dedication and the location where the altar was found, it seems not unreasonable to suspect that Celer was perhaps a naval officer of the *Classis Pannonica*, the fleet formed to operate on the upper half of the Danube and whose headquarters were at Vindobona. Romermuseum, Vienna. (*Author's photographs*)

Neptune, the god of the sea, was an obvious subject for veneration by sailors and altars dedicated to him have been found all around the Roman world (including on the Danube) usually on votives either praying for or giving thanks for calm seas on a voyage. The more ethereal Oceanus, the deity of the great river which was thought to surround the earth, was although less commonly, likewise a figure to be placated.

With the addition of Egypt to the Empire after 30 BC, many Egyptian seamen joined the navy and brought with them the goddess Isis. She was a very ancient Egyptian deity, part of their trinity of Osiris, Isis and Horus and who had become the most important of their goddesses.[7] She had been further adopted by Ptolemy I Soter (reigned 305–282 BC)[8] and given a consort, Serapis, a synthetic deity made up of Egyptian and Greek elements, to

83

Also popular among seafarers as a protective deity, was the sea-nymph Thetis, left, daughter of the sea god Nereus, here shown holding a ship's rudder. First century AD. Metropolitan Museum of Art, New York.

Above right, Isis. This ancient Egyptian deity was adopted by seamen as 'the star of the sea' and their protectress. Early third century BC. Archaeological Museum, Dion, Greece. (*Author's photographs*)

give a religious focus for his assumption of pharaonic power.[9] She had in her own right, always remained a popular, fundamental and supreme figure, a status that was continued, in her manifestation as 'the Star of the Sea' and thus revered by seamen as a maternal figure. Similar in nature and identifiable for sailors from the Levant, was the Phoenician goddess Asterat of the Sea (Astarte in Greek) a deity that was interchangeable with Isis in her appeal to sailors. Isis had a widespread following, even penetrating to beyond the empire, into Germany, where among the Suebi her emblem was a 'light warship', once more acknowledging her connection with the sea.[10] The spring festival that marked the opening of the annual sailing season became dedicated to Isis as protectress of shipping, a tradition that was carried over into Christian times, substituting the Virgin Mary and which continues to this day.[11]

Another important religious cult was that of Mithras, who represented celestial light, manifested by *sol invictus* or the unconquerable sun, often depicted in a tauroctony, the figure of Mithras slaying a sacrificial bull.[12] The belligerent attitudes demonstrated by this act, allied to the doctrines of an immortal souls and good moral conduct, together with fraternal help among followers (women were barred) appealed strongly to the soldiery of the empire. Many a Mithraeum has been found linked to military bases, dating particularly from the first and second centuries AD. Widespread in the army, it is not known to what extent the cult spread amongst navy personnel, but some indication that it was widespread can be gauged from an altar or statue base dedicated to Mithras in the mid-third century AD by none less than the prefect of the Misene fleet, the second most senior military officer of the whole empire.

The navy also subscribed in imperial times, to the cult of the deified emperors, the imperial

The official imperial cult of the deified emperors was followed by the navy, as well as everyone else. This photograph shows the ruins, adjacent to the western side of the outer basin at Misenum, of the temple complex dedicated to the cult of the emperors, in this case, the Flavians.

The reconstructed facade of the temple, with statues of Vespasian (left) and Titus (right); there is also a bronze equestrian statue of Domitian (out of shot, on the right) altered by the substitution of the head by that of his successor, Nerva. The pediment includes portraits of Antoninus and Faustina. Mid-second century AD, now in the Castle Museum, Baia. (*Author's photographs*)

cult, as witness the temple of the Augustales at Misenum, dedicated to the Flavian emperors. The growth of Christianity, until it became the official state religion of the empire, must be assumed to have spread covertly among navy personnel in the same way that it did in other parts of society, despite proscription and persecution. Eventually of course, the whole Roman world embraced the new religion, however sailors confronted by such changes, while the sea remained unchanging doubtless stuck to their traditions, festivals dedicated to Isis continuing into the fourth century AD and has been seen, the tradition being so deeply ingrained that it had to be adopted and adapted for Christianity. They still paint eyes on the prows of boats.

PART VI   SEAMANSHIP

16   NAVIGATION

BY THE TIME of the founding of Rome in the eighth century BC, seamen had long since sailed, explored and become familiar with the basic geographical and weather features of the Mediterranean Sea. They knew that the prevailing winds of summer mostly blew from Europe towards Africa (with occasional winds in the opposite direction); that there is a current from the influx of the Atlantic Ocean through the Straits of Gibraltar (Pillars of Hercules) which travels in an anti-clockwise direction along the North African coast, with a loop around the Gulf of Sirte, on to the Levant, around the Aegean and Adriatic and northward up the western Italian coast and on, via France, to Spain; that tides are virtually non-existent, save in a few restricted waters.[1] The prevailing seasonal winds of the Mediterranean are quite regular, leading the ancients to formalise and to name them as emanating from twelve points of the compass (so to speak) any wind from an odd direction being regarded as an amalgam of two or more of the named winds.[2]

In ancient times and indeed through to medieval times in the Mediterranean, there were certain months of the year when navigation of the seas was undertaken and other, winter months when it was not, namely sailing and closed seasons 'for the violence and roughness of the sea does not permit navigation all the year around.'[3] Apart from the weather, the other reason was that, with navigation so reliant upon the ability to make observations, winter visibility, mist and fog, together with the short hours of daylight, seriously impeded the ability to do so. The custom became therefore that the sailing season was from March to October inclusive and the opening of the annual season was cause for celebrations.[4] Vegetius, writing in the fourth century AD goes further, saying that for 'fleets' the season should be from 27 May to 14 September.[5] That is not to say that voyages

were not made in the 'closed' months; Caesar's crossing of the Adriatic in pursuit of Pompeius, with seven legions, was made in January[6] and Saint Paul was famously shipwrecked on Malta, his ship having sailed after the close of the season for that year.[7] Conversely, voyaging during the 'open' season did not guarantee fair weather as evidenced by the losses to Roman fleets from storms during the First Punic War. The custom of a sailing season was continued by the Romans when they started to operate on the northern coasts of Europe.[8]

The earliest sea navigation was effected by following coastlines, for obvious reasons which remain valid to this day, but which were particularly relevant to the operational limitations of ancient warships. As has been seen they were evolved for speed and power in fighting their own kind and their design precluded good sea-keeping ability. They could not weather heavy seas, mount effective blockades or stay at sea for more than a very few days at a time. Voyages were preferably made by stages along coasts and between regular resting and watering stops, a practice aided by the fact that within a dozen or so miles of land, during the Mediterranean summer, the winds are mostly either onshore or offshore caused by the fact that the landmass heats and cools far more rapidly than the sea. These breezes mostly override the prevailing winds, found further offshore.[9] Even taking a coastal route could however prove risky due to the suddenness with which adverse weather can arise close in to shore. In AD 64 Nero ordered a squadron of the Misene fleet back to base regardless of weather and somewhat late in the season; they had to travel only the short distance from Formia (Formiae) but the squadron was caught in a south-westerly gale and several warships and smaller craft were wrecked actually on Cape Miseno, within a few yards of the harbour mouth itself.[10] They had been the victim of a sudden squall, caused by the winds dropping rapidly down the sides of the coastal mountains on to the sea, its strength and speed accentuated by the sudden contrast in temperature between the land, baked by the sun, and the cooler sea.[11] Open sea crossings, such as between Sicily and Carthage or across the Black Sea were undertaken when conditions were favourable and the journeys of known distance and direction. Ancient naval battles, however, all took place in sight of land, the local coast being more often than not used as a factor in the battle and a possible salvation for stricken ships and survivors in the water.

The ancient Egyptians had conjectured from seeing sails disappear over the horizon, that the Earth was 'curved'. The Sumerians had by 2100 BC calculated the length of a year at 360 days, based upon the lunar cycle. This figure, although corrected, has been retained and adopted as the number of degrees of subdivision of astronomical and navigational observation ever since.[12] The Babylonians divided the day into twenty-four hours, which the Romans halved to twelve hours of day and twelve of night, irrespective of the varying length of daylight through the year, enabling sundials to be calibrated accordingly.

Navigational aids and methods

The Phoenicians had navigated from observing the constellations, especially the Great Bear or Plough. The Pole Star, presently fifty-eight minutes off true north, was further away from it in ancient times, an error known to the ancients; Hipparchos of Bythinia, in the mid-second century BC, noting and commenting that there were no stars directly above true north, only a void with some stars close by, including Stella Polaris. The earliest maps to show cardinal points date from about 2300 BC Mesopotamia[13] and scales of distances on maps, from Babylon. Further observation and consideration of astronomical phenomena led to the concept of latitude and that this could be used to determine position relative to north and south. Pytheas of Marseilles in about 320 BC, travelled across France and beyond

the north of the British Isles but importantly, made astronomical observations throughout his journey.[14] Hipparchus codified a method of describing the position of places by reference to imaginary lines stretching from north to south (latitude) and from east to west (longitude). He also compiled a star catalogue for night navigation, although the Egyptians had, considerably earlier, already made star-clock charts to calculate the time at night according to the rising and setting of certain stars. Shortly after, one Marinus of Tyre regularised the spacing of the grid lines; he also sought to solve the problem of projecting the spherical nature of the earth on to a flat map, a problem greatly improved by the geographer Claudius Ptolemy in the second century AD.[15]

Building upon the work of the Sumerians, Egyptians and Babylonians, Greek astronomers formed a model of the Solar System and stars to explain their relative movements although the predominant view was that it all revolved around the Earth. Eratosthenes of Alexandria (276–194 BC) deduced the tilt of the Earth's axis to within one tenth of a degree, thus explaining the changing of the seasons.[16] He also calculated the Earth's circumference at 24,700 miles (39,742 km) only 202 miles (325 km) less than the actual figure. Although at most nominal in the Mediterranean and Black Seas, the ebb and flow of tides was known and understood, as was their cause, as Vegetius says, 'since in different regions it [the tide] varies at appointed times according to a different state of the Moon's waxing and waning,'[17] confirming similar observations by Pytheas and Pliny.

Although, as far as we know, they did not have charts as we know them, ancient mariners were well served by publications containing sailing directions and details of coasts, routes or bearings and harbours, to assist in the basic navigation by landmarks and stars. In the mid-fourth century BC, a geographer named Scylax produced a *periplous* (literally, a sailing around) which described coastlines around the Mediterranean with details of ports, rivers and places where fresh water could be found.[18] Another one that has survived from the mid-first

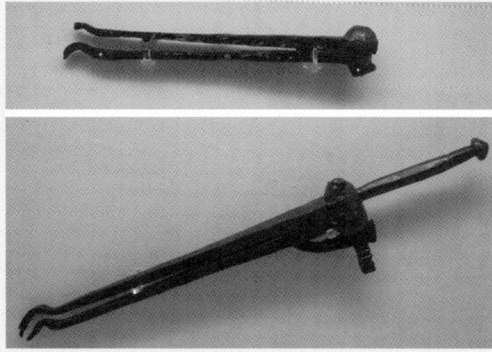

Instruments for navigation. Above, iron dividers or compasses, second or third century AD, found at Lyons (Lugdunum). The lower set can be locked so that measurements can be transferred. Lyons Archaeological Museum.

Right, replica of an instrument from Phillipoi in Greece, third century BC. A precision instrument used to calculate, latitude and the height and angular distance above the horizon of the sun or other stars. Thessaloniki Archaeological Museum. (*Author's photographs*)

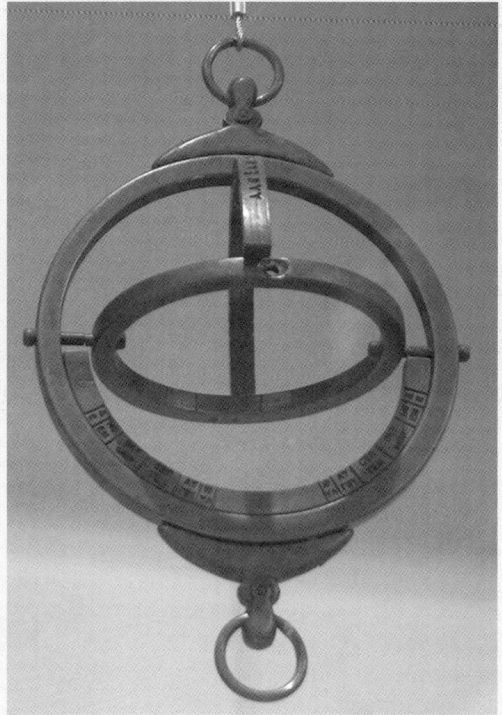

century AD is known by its Latin title, *Periplus Maris Erythraei*. The Erythraean, or Red, sea, extended beyond what we now call the Red Sea to include the Gulf of Aden, the Persian Gulf, parts of the Indian Ocean and routes to India. It also detailed the Indian coast from the Indus to the Ganges and the East African coast to Dar es Salaam.[19] Another such coasting pilot is known to have been for a circuit of the Black Sea, prepared for an intended voyage by Hadrian. Schedules of sailing directions were compiled and available for voyages between specified points, including estimates of journey times according to given wind conditions. Thus, a route would be described as for example, 'from Pylos (western Peloponnese) in fair winds, sailing due west for four days at an average three knots will bring landfall on eastern Sicily.' It could as well be expressed as a journey along a given line of latitude to a destination bisected by that line, or by sailing at a given bearing from a certain point to a destination in that direction. For Aegean crossings, island hopping

was and is the simplest method of navigation. Cicero described his own journey in 51 BC from Athens to Ephesus in a warship, with stops at Kea, Kythnos, Siros, Delos and Samos on the way, a leisurely voyage that took two weeks instead of the normal three days.[20]

Instruments to assist the taking of accurate observations for navigation were developed. Sundials as a means of telling time date from at least 1300 BC, the date of a set of instructions for making one, found in the tomb of the Pharoah Seti I.[21] Water clocks or clepsydra also date from about the same time in Egypt and became common in the ancient world. Instrument making had become refined by the third century BC, as evidenced by a form of astrolabe from Greece, by which time, latitude and the height of celestial bodies could be calculated. From the first century BC, comes the Antikythera computer, an intricate mechanical device of a series of twenty-two toothed, meshing gear wheels which, upon setting the date, yielded the relative positions of the sun, moon and stars.

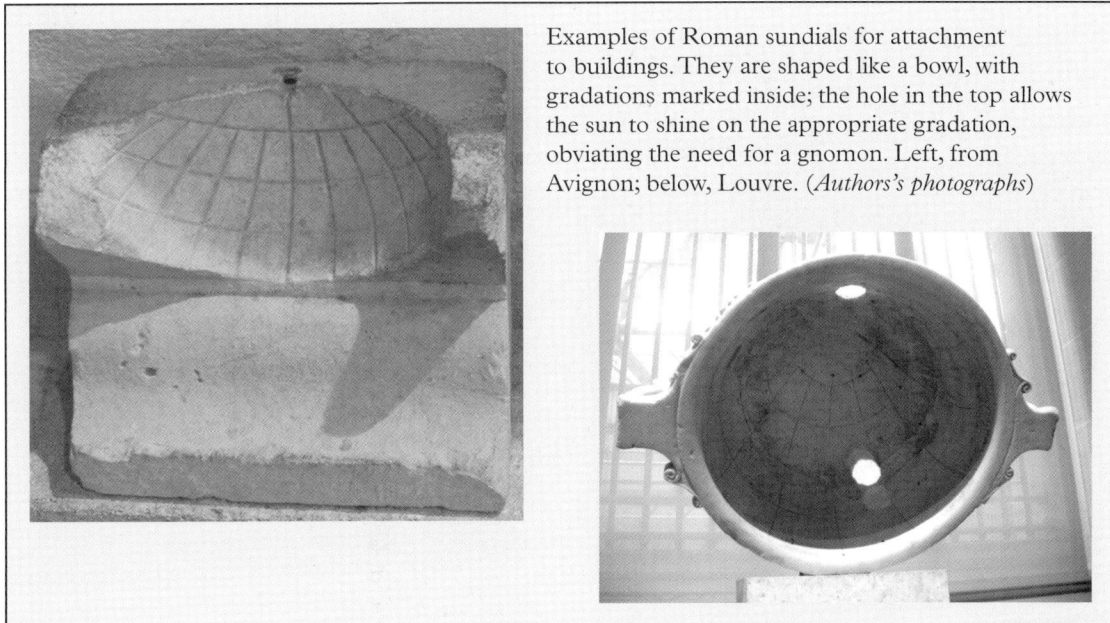

Examples of Roman sundials for attachment to buildings. They are shaped like a bowl, with gradations marked inside; the hole in the top allows the sun to shine on the appropriate gradation, obviating the need for a gnomon. Left, from Avignon; below, Louvre. (*Authors's photographs*)

Comparison of the readings with observations would then allow position to be found.[22] These devices, together with the Roman portable sundial and Hipparchos' and Eratosthenes' calculations worked because it had been established that the angle of a celestial body from the observer's zenith, i.e. an imaginary vertical line extended through and above the observer, was the same as the angle between the centre of the Earth and the point upon its surface directly below the celestial body, i.e. its zenith. This meant that measurement of e.g. the Sun's angle at midday, adjusted for date and deducted from ninety degrees would yield the observer's latitude, a function still performed by a modern sextant.[23]

The Romans developed a small, portable version of the sundial, of added sophistication. Normally fixed in a sunny position with its *gnomon* (pointer) aligned north and south, the shadow of the sun falling across the scale indicates the time. The Romans added centrally pivoted discs or rings calibrated as to latitude and as to declination of the sun. The former they calibrated at between thirty and sixty degrees north, the whole of the Roman Empire

laying between those latitudes. The latter they had calculated by using the number of degrees difference between the angle of the sun at midday on the equinoxes and the solstices. Eratosthenes had found that marking the equinoxes as zero, the maximum difference was twenty-three and a half degrees south (the winter solstice) and the same amount north (the summer solstice) further proving that the Earth is round. Used in the conventional way therefore, the device told the time; knowing the time and latitude, pointing at the sun and reading off the declination to correct seasonal variations, would give true north.[24]

Longitude, that is the distance east or west of a particular point, remained an inexact matter until Harrison's chronometer of 1762.[25] The only way was to keep records of time and estimated speed through the water, to calculate an estimate of the distance covered and by factoring in bearing and latitude, to triangulate an approximate position. For open sea passages between known points, the sailing directions served to reduce the amount of guesswork needed and for warships following coastal routes, the coastal pilots were all that was needed.

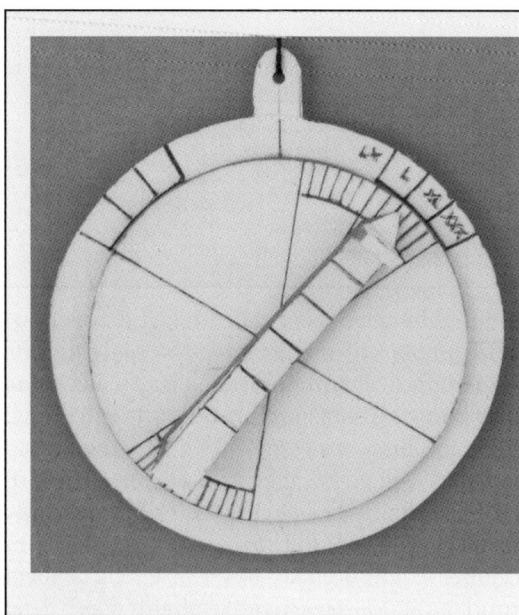

Author's model of a Roman portable sun compass and sundial. It is graduated on the outer ring as to latitude (between 30 and 60 degrees north). The inner ring is calibrated as to the declination of the sun, equivalent to the date, divided into fifteen-day periods. The *gnomon* is divided into six sections; the Roman day was divided into twelve hours of daylight, six before and six after midday.

To use as a compass the latitude, date and time must be known. The centre line of the inner disc is set against the latitude on the outer. The south end of the *gnomon* is set against the declination (date) on the inner scale; knowing the time, the disc is rotated until the sun's shadow falls onto the appropriate mark along the *gnomon* and when the device will be aligned to true north and south.

With the device aligned to north and south and with the latitude and declination indicators duly set, it acts as a conventional sundial.

Lighthouses and canals

The first recorded lighthouse was the famous Pharos of Alexandria, built about 270 BC on the island of the same name. It was about 350 feet high (107 m) in three storeys; the lowest was square, about 180 feet high (55 m) and housed the crew. Atop this was an octagonal section about 90 feet high (27.5 m) and above this in turn a 60-foot high (18.3 m) third storey, round in section. On the very top was a cupola with eight columns in which was housed the lamp, a brazier burning resinous wood, the range of its light enhanced by a system of polished metal convex mirrors.[26] It remained in service throughout the Roman period and set the pattern for the system of lighthouses that the Romans developed to guide sailors, throughout the Empire.

The largest, surviving example is still in service at Corunna in north-west Spain. It was originally built under Trajan in the early second century AD. Another large lighthouse started life as a monument erected by Gaius at Boulogne, but was later converted; it was some 200 feet high (61 m). Across the Channel, on the heights at Dover were two lighthouses, one each side of the harbour and dating from about AD 100; one survives to about half of its estimated original height of about 80 feet (24.4 m).[27] Most major harbours had lighthouses, including Misenum, Ostia and Fréjus (Forum Iulii). All were of several storeys (Ostia had seven, Boulogne had twelve), round or polygonal and although none were as large as Alexandria, thirty or so have been identified as being in service around the Empire by AD 400.[28]

The excavation of canals was another activity undertaken by the Romans to aid navigation and communications. The oldest canal within their remit was the canal dug in about 610 BC for the Pharoah Necho II, linking the (since dried up) Pelusiac branch of the Nile Delta, via Lake Timsah and the Bitter Lakes, to the Red Sea, a route of 60 miles (100 km).[29] The canal sections were extremely prone to silting, which reduced its usefulness and even closed it from time to time. Augustus had some restoration work carried out when he took over Egypt and it was probably at least partially usable for the expedition that he sent down the Red Sea in 26 BC.[30] The Romans did not normally maintain naval forces in the Red Sea, only sending ships as the need arose. In AD 106, Trajan annexed the kingdom of Nabatea and in preparation for this, he had the canal restored and dredged and sent a fleet through it to attack the Arabian coast.

Above, mosaic showing a lighthouse; the inscription says 'ho pharos', the lighthouse, in Greek, thought to be that of Alexandria,; second or third century AD. Qasr, Libya. (*Photograph, Pat Aitchison*)

The lighthouse at Corunna (ancient Brigantium) in north-west Spain. Originally built under Trajan (AD 98–117) although it has been renovated and the exterior re-clad, the core remains original; it also bears a more than passing resemblance to reconstructions of the Pharos of Alexandria. To invest in such a building at this location suggests the existence of an active maritime trade route, passing along the Atlantic seaboard in Roman times. (*Author's photograph*)

Thereafter the responsibility for the canal was handed to the *Classis Alexandrina* to maintain and patrol and for any forays they might make into the Red Sea. Hadrian was thus able to use the canal, when he sent a squadron to deal with pirates in the Red Sea.

Nero's reign saw several schemes for canals, first to join the port of Pozzuoli to Rome across the Pontine Marshes, a work in progress that died with the emperor.[31] Another, very practical scheme was to link the rivers Moselle and Saône in Gaul, which would have joined the Rhine to the Rhone and permitted water-borne transport from the Mediterranean to the North Sea. Despite the obvious strategic advantage of this, the scheme was thwarted by local opposition by those employed in the portage trade between the rivers.[32]

Nero's final scheme was to dig a canal across the Isthmus of Corinth. Work was started in AD 67 but ceased when Nero had to leave for Rome.[33] The reign did see some canals completed and after suffering shipping losses off the Belgian and Dutch coasts, a 23-mile (37 km) canal was dug to link the Rivers Meuse and Rhine.[34] In Britain by the late first century AD, a system of canals had been dug linking the rivers and inlets and enabling year-round water-borne transport between London (Londinium) and the legionary base at York (Eboracum) without risking the open sea.[35]

17 SHIP PERFORMANCE

ASSESSING THE PERFORMANCE of ancient warships is made difficult by the lack of contemporary data or of any more recent comparable ships, with the sole exception of the trireme *Olympias*. For monoremes it is possible to make some approximate comparisons with modern rowing boats, but it is when considering warships configured as biremes and triremes that the lack of data becomes a problem. What is available is a large number of ancient accounts of voyages by ships, fleets and even of differing ship types from which at least average speeds can be calculated. For warships however, simply having the journey time and distance does not serve to differentiate between progress made under oar, sail or a combination of both.

Given that the *Olympias* is an accurate replica of a fifth century BC Athenian trireme, the results of her sea trials provide an excellent starting point, with actual figures, to estimate, or more accurately, make reasonable guesses at, the capabilities of say a first or second century AD trireme (the last triremes appeared in action at Hellespont in AD 323). However this data must perforce also be used as the yardstick to estimate performance for all of the other bireme and trireme forms of ancient warship. In sea trials under oars alone, *Olympias* attained a speed of 8.3 knots in a short burst and peaked at 9 knots; 7 knots was maintained over a nautical mile and 4.6 knots[1] for 31 miles with the crew rowing in shifts.[2] This compares well with US Navy light twelve-oared 31-foot (10 m) racing cutters that could maintain 7 knots in ideal conditions over 3 miles (5 km).[3] Maximum striking rate for the oars was thirty-eight strokes per minute;[4] it is interesting to compare this with the last (2011) Varsity Boat Race, where the striking rate averaged thirty-four strokes per minute, peaking at thirty-seven, but this was on a calm river at near horizontal, on sliding seats allowing a much longer stroke (yielding a burst speed

of 14.74 miles per hour (12.8 knots, 23 kmh). Lessons learned from the operation of *Olympias* have indicated that proposed amendments of the design and internal arrangements would improve these figures but even without any such, the ship was found to be very handy, able to stop in one length and with a turn rate of three degrees per second, had a turning circle of only 3.4 lengths at 7 knots; with one side oars stopped, she could turn in only 1.9 lengths with a fifty percent drop in speed, in two minutes. Under combined oar and sail she made 6.6 knots with a 20-knot tailwind. Under sail alone, *Olympias* was stable in winds of 25 knots, rolling ten to twelve degrees; in a 20 knot wind, she managed 10.8 knots and cruised comfortably at 7 knots.[5]

Also known are the capabilities of some of the replicas of the later Saxon and Viking ships which have been built and sailed, particularly replicas of the early eleventh-century AD Viking ships from Skuldelev in Denmark.[6] The two warships from there (nos 2 and 5) recorded speeds under oar of 3.4 and 5.4 knots and under sail of 13.8 and 15 knots respectively. The replica ninth-century AD Viking Gokstad ship, sailing to America in 1893, was recorded at 11 knots in the Atlantic.[7] Obviously, these ships are 500 and more years later but, so far as the barbarian craft are concerned, their basic configuration is fundamentally unchanged, although considerably refined, over the ships of the period under consideration and as such, the performance of these much later ships represent a substantive improvement. From all of the trials it seems that sailing performance was broadly similar although the Saxon and Norse ships could possibly withstand more extreme seas than the trireme. It can only be a possibility because although the tenderness of a Mediterranean trireme is attested, those evolved for northern seas were more suited to them; further, there is no information as to the numbers of Roman and barbarian craft that foundered in heavy weather or were lost in storms there. Under oars however the trireme has a clear advantage both as to speed and power and would within reason, have little difficulty in catching the rest, especially if they had no rig as seems likely in the early days of barbarian raiding.

The foregoing could well explain how barbarian craft, caught in the Aegean and eastern Mediterranean by Roman warships rarely made it home. In northern waters it can be assumed that Roman ships benefitted from the long Celtic tradition of navigating those seas and were of more substantial build with higher freeboard than their Mediterranean forebears, to make them more weatherly.[8] This would be likely to reduce their speed under oars a little, but improve their handling and sailing capability. Overall it is felt that Roman warships in the north would have comparable, if not slightly better sailing performance, given their much longer experience and better sailcloth, than their barbarian enemies and be faster by a significant margin under oars.[9] Bigger, faster and better armed, the Roman ships were superior; when linked to the shore-fort system of interlinked watchtowers, forts and naval units, the system successfully kept barbarian activity at bay or at worst, minimised, so that the provinces so protected, Britain in particular, continued to prosper. This success depended on the system being kept fully manned, well maintained, well trained and in a state of constant readiness and the position quickly deteriorated when this was not so.

Examples of voyages

There are many accounts of voyages by ancient warships which can give an appreciation of the lengths of voyages actually undertaken and relative performance of the ships under cruising conditions. There are of course, variables of wind and sea state to be taken into consideration, together with the mode of propulsion, whether by oars alone in total, or in relays,[10] sail alone or a combination of both. Voyage times could vary

greatly so that, as an example, the much-crossed Sicily–Tunisia route of approximately 110 miles (176 km) was achieved in 204 BC by Scipio's invasion fleet of warships and transports in about thirty hours, an average of three knots.[11] In 49 BC, a smaller fleet again of warships and transports took sixty hours for the only slightly longer journey to a point near Nabeul.[12] Two years later, Caesar's crossing with an invasion fleet took three and a half days, presumably having suffered worse conditions.[13]

Warships alone could make some prodigious journeys given favourable conditions, thus a fleet of quinqueremes crossed the Adriatic from Brindisi to Corfu, a distance again of 110 miles (176 km) in only eleven hours, making 10 knots, obviously with a following wind.[14] Conversely, a fleet of Athenian triremes, under oar alone, could only manage 31 miles (50 km) in twelve hours, a meagre two and a half knots, in poor conditions.[15] Where possible, ships would proceed in stages, between stops for rest and food. Thus a 'long day' journey from Byzantium (Istanbul) to Heraclea (on the north Turkish coast) in 400 BC by Greek triremes covered 129 miles; after the first 64 miles (124 km) they stopped for a meal ashore, before travelling another 62 miles (112 km). The 'long day' must have been of over eighteen hours, plus the time spent at the stop, in order to average seven knots, the cruising rate achieved by *Olympias*.[16]

Within a similar compass, the Roman naval base at Chersonesus (near Sevastopol, Crimea) was within a one-day voyage of the Danube Delta (120 miles; 200 km) and Sinope on the north Turkish coast (100 miles; 160 km). Both are well within the non-stop voyage by triremes recorded between Athens and Mytilene, of 184 miles (340 km).[17]

Endurance

Cicero's leisurely voyage across the Aegean, previously mentioned, was not typical although demonstrating that warships could, with regular stops, cover long distances.[18] Thus a Roman fleet of about 120 warships was despatched from Ostia in 217 BC to hunt a Punic fleet of seventy ships which had been reported near Corsica and Sardinia and which had, off Pisa captured a Roman supply convoy bound for Spain. The Punic fleet fled upon the Romans' approach and were then pursued by them around Corsica The fleet finally returned to Ostia.[19]

The cruise of this fleet, predominantly of quinqueremes, can serve to demonstrate the endurance achievable, given the fair weather that they obviously enjoyed. From examples of trireme cruising voyages, an average distance of about 100 miles (160 km) per day could be covered. These ships were bigger, heavier and not quite as fast, so a daily average of approximately 80 miles (128 km) is acceptable. Starting from Ostia, it would have taken three days to cover the 212 miles (339 km) to reach Pisa (then a sea port) with stops say at Argentario and Piombino. Assuming that the ships carried three days' rations (see below) they had to draw stores at Pisa. The crossing to Bastia on Corsica (80 miles, 128 km) took another day and from there along the east side of Corsica to Olbia (160 miles, 256 km) implies another overnight stop, reprovisioning at Olbia on Sardinia. Two more days (with an overnight stop) brought them to Cagliari, Sardinia (173 miles, 277 km). The next stage of the voyage required an open sea crossing of some 240 miles (384 km), a three day non-stop journey to Trapani on Sicily, where they would have again revictualled, rested and cleaned ship. The short voyage to Marsala (15 miles, 24 km) then their crossing to Africa and return to Sicily took another three days. Finally, they returned to Ostia, another eight days of voyaging. The fleet had been cruising for at least twenty-three days and had covered some 1,790 miles (2,864 km) in that time (even further if they had gone west-about Corsica and Sardinia). The fleet repeated the cruise in the following year, 216 BC.

and Sardinia, across to western Sicily; there the Roman fleet paused before crossing to Africa to ensure that the enemy had indeed gone home, taking the opportunity to raid the enemy coast.

On another occasion, in 210 BC, a Roman fleet sailed from Sicily on an extended raiding operation against the Carthaginian coast of Africa. The fleet raided extensively and must have subsisted largely on plundered provisions, only returning to base at Marsala after a thirteen-day cruise.[20] The three-day voyage period recurs in these accounts and also appears for example, in AD 467, when a Roman war fleet sailed from Cagliari to the African coast, taking three days.[21] As previously posited, it would appear to confirm that, bearing in mind the very limited stowage available on ancient warships and the necessity of devoting most of that to the carrying of water, that ships normally carried three days' rations as an operational norm.

Relative speeds

Although there is evidence for the performance of an Athenian-type trireme, there is none for the other types of warship in Roman service. It is generally accepted that the trireme was the fastest of the ancient warship types and thus able to catch anything smaller and to escape anything bigger.[22] This advantage could be lost to the better seakeeping ability of a larger ship in heavier weather or against strong currents.[23] In equal calm conditions, it has been estimated that the quadriremes, quinqueremes and sexteres were probably one knot slower than the trireme on average, with ships larger and smaller than these slower by one, to one and one half knots, the former due to their sheer bulk and the latter because of their smaller and less powerful oar systems.[24]

The Republic

Before the First Punic War, the Roman navy was of modest size and the few warships employed by Rome and her naval allies had to operate from and share the commercial ports. There is no indication of the existence of any dedicated naval ports or solely naval sections within existing harbours, ships would use port facilities and acquire provisions from local merchants; there must have been retained at principal ports perhaps a warehouse or store for warship gear, spare rigging and such like, especially after the appointment of the 'navy board' in 311 BC.

With the huge growth of the service for the Punic Wars, the pressure and conflict between the much-expanded navy and the merchant marines for harbour space must have been intense but there is still no record of separate naval bases in that period; perhaps the problems were alleviated by the reduction in civilian traffic due to the Wars and the way in which fleets were constantly operating in different places.

As to the nature of supporting shore facilities in the most frequently used ports, such as Ostia, Naples and Palermo, the increased demands of the navy required the obtaining of premises, camps or barracks, for housing, feeding and equipping recruits during training, the same for off-duty crews. As an indication of the scale of the problem that victualling officers had to address, fleet of 100 quinqueremes had crews totalling about 37,000 men, more than the population of most of the towns that they visited. Also needed were premises of commensurate size for storing and repairing the far larger quantities of gear of all sorts, as well as facilities for amassing and keeping ship's stores and victuals in the large amounts now necessary to supply and outfit an operational fleet, frequently of over 100 ships.

Given the particular nature of warships, it is probable that certain shipbuilders had come to specialise in their building. In the build-up to the wars, with the expansion of the navy, additional shipyards were required and although these

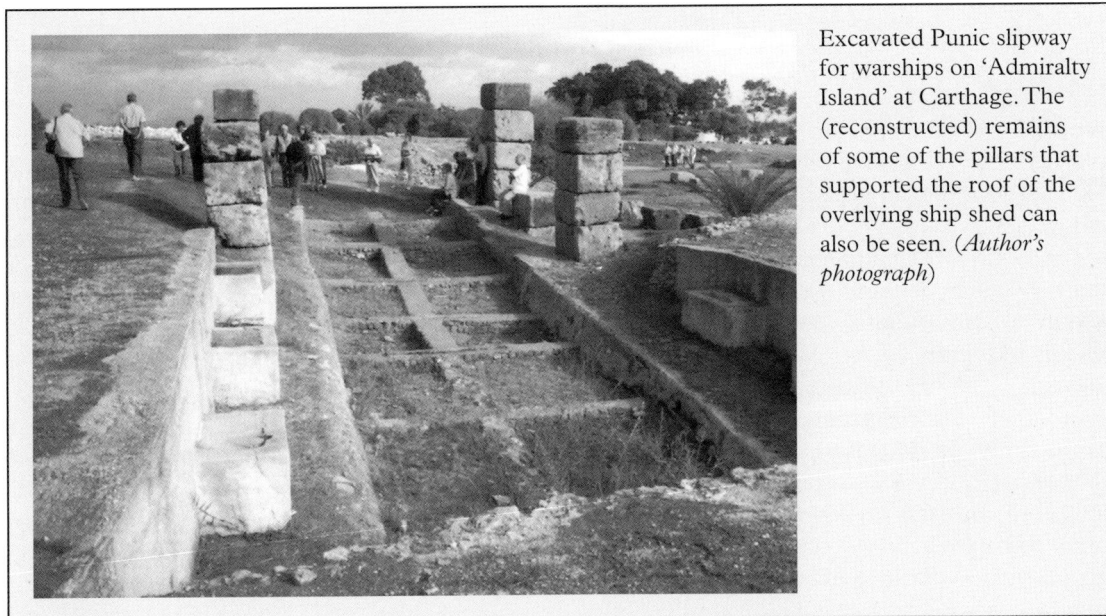

Excavated Punic slipway for warships on 'Admiralty Island' at Carthage. The (reconstructed) remains of some of the pillars that supported the roof of the overlying ship shed can also be seen. (*Author's photograph*)

The site of the naval base of Misenum, seen from Monte Procida. The greatest naval base of the Imperial navy and headquarters of its senior fleet, the *Classis Misenensis*. The harbour is formed from two flooded volcanic craters, the outer protected by moles. Little now remains of what was once a massive installation with shipyards, jetties and piers, shipsheds, workshops, storehouses and victualling depots. Additionally, there were dozens of warships and thousands of men, together with the town that surrounded it all. The fleet prefect's house was on the Cape, facing the camera, with the base surrounding the basins and the civilian town in the foreground and to the left. The great cistern is within the low hill to the left of the inner basin.

The outer basin of the harbour at Misenum, looking east towards Vesuvius and where one of the great harbour moles once enclosed it. This rather sleepy, pleasant pleasure harbour is a far cry from the great hive of activity of its days as a naval base. The topography of the whole area had altered greatly since the Empire owing to volcanic activity. (*Author's photographs*)

yards were privately owned, as a specialised activity the building must have been carried out under naval supervision by officers, known from the imperial period as *fabri navales* (naval constructors) and *architecti* (naval architects). As the wars progressed, this became more necessary with the constant demand for new ships and the overhaul and repair of serving ships. During the Second Punic War certainly, there was a programme whereby combat ships were withdrawn in rotation from the fleets for regular refitting; thus in the commissioning of the fleet for 208 BC, thirty of the ships were sent for refit[1] and again, in 204 BC, another thirty were in refit.[2]

Shipyards are assumed to have been concentrated at or adjacent to the main ports such as Ostia and Naples to be close to collection points for materials and populations that could provide workers. For premises for building and launching the ships, a suitable beach would suffice and sheds for workshops and storage could be quickly added and jigs and cradles, even for the largest of ships were all of wood, none of which unfortunately leaves remains. With the end of the wars and the reduced requirement for new ships, the 'wartime only' yards quickly disappeared or converted to building merchant ships. There are the remains of permanent stone slipways at Thurii in Calabria and shipsheds for warships at various places (such as Athens and Carthage) with inclined ramps for ship launching survive and shipyard installations would have been similar.

It is during the civil war period and the war between Octavian and Sextus Pompeius from 38 to 36 BC, that the first recorded dedicated naval facility was set up by Octavian's admiral, Agrippa. The base was established by digging canals to link two flooded volcanic craters near Pozzuoli on the Bay of Naples, Lakes Averno (Avernus) and Lucrino (Lucrinus) to each other and to the sea. The base was named Portus Iulius and the lakes formed secure lagoons in which to train crews and house the ships. Extra channels were dug to allow a flow of seawater to prevent silting and tunnels 10 feet (3 m) wide were bored through hills behind the base to link it to Naples and Cumae and the routes to Rome. A subsidiary base was set up at Fréjus in France with enlarged and improved, fortified harbour facilities.

Imperial fleets

With Augustus' reorganisation of the navy into several permanent formations, came the establishment of permanent, dedicated naval bases to house them. Each fleet was a fully self-contained entity, a navy on its own with every necessity for its operations.[3] Each had a headquarters base with sub-bases at various points within its operating area. The main base would of course house the fleet prefect with his staff and administration, together with a garrison of marines and the crews of the ships in port and stationed there. There were quarters and barracks to house them, baths, bakeries and kitchens, together with a base hospital. In

addition there were slipways, workshops and at the big bases such as Misenum, full shipbuilding capability. All had the shore-based personnel, craftsmen and workers who refitted and maintained the ships and made and repaired their gear and fittings, as well as armourers for the weapons and armour. Additional personnel were responsible for food and stores, victuals, weapons, clothing so that just about every trade imaginable must have been on the strength, employing thousands of men. Around the bases towns quickly grew up.

Subsidiary bases could be anything from a small store with a few men to ensure that visiting ships could be looked after, to complete miniatures of the main fleet base with a garrison and all necessary facilities to support operations and having hundreds more men. The two Italian Fleets each had a detachment of sailors stationed in Rome itself, at first billeted in the Praetorian Barracks and later moving to their own quarters, the *castra Misenensis* near the Colosseum and the *castra Ravennate* on the West bank of the river. Their role was to provide couriers for the transmission of dispatches to and from the Fleets but they were also in charge of the awnings of sailcloth that were rigged and extended across the top of the Colosseum to provide shade from the sun. Another task for them was to organise and run *naumachiae* (mock sea-fights) between ships manned and fought by the condemned and staged in flooded arenas.

The Romans, certainly under the Empire and most probably well before that, gave much effort to the organising and provision of medical services and facilities for the fighting men (to an extent in fact that would not be seen again until the nineteenth century AD). Naval bases had base hospitals, such facilities having been identified at bases on the Rhine and Danube.

The bases

The principal permanent naval base was set up at Misenum at the northern tip of the Bay of Naples, from 22 BC. Despite their efforts, Lake Lucrino silted badly and became unusable. The new location again made use of flooded

PART VII   OPERATIONS

19   STRATEGY AND DEPLOYMENT

The Roman world-view

For the first few centuries of its existence Rome's attention was focused on and limited to her immediate surroundings. This view expanded with the growth of Roman and allied territory and with it the need to look beyond to the wider world. That is not to say that the Romans were ignorant of what lay beyond their immediate ken; the city was always a trading centre and her returning merchants as well as those visiting from elsewhere to trade brought such knowledge. Contacts with Carthage revealed lands to the south and west, with the Greek lands to the east and from the Etruscans and Gauls they learned of those to the north. That Roman merchants could and did venture abroad from the earliest of times is indicated by Rome's first treaty with Carthage in 509 BC, which forbade Roman ships from entering African waters west of Cape Bon,[1] the Carthaginians finding it necessary to restrict Roman trading voyaging.

In the fourth century BC the conquests of Alexander the Great brought to the Mediterranean World even more detailed knowledge of lands far into Central Asia and India. Trading stations were later established in India and even far China became the subject of an embassy sent in the reign of Marcus Aurelius.[2] In about 320 BC Pytheas of Marseilles returned from his epic journey far to the north of Europe and wrote an account of his travels, *On the Ocean* (since lost), across Gaul to Britain and beyond.[3] In addition early Phoenician mariners had brought knowledge of the coasts of West Africa, probably the Canary Islands and the sea routes around the Iberian Peninsula and Gaul to Britain.[4]

There were other geographers and writers providing accounts of foreign travel and of lands beyond Italy, whose writings were known, such as Herodotus (writing in the mid-fifth century BC) and Xenophon (early fourth century BC). Later, Caesar's own memoirs (later first century BC) together with the writings of such as Pliny and Tacitus (late first century AD) were to be supplemented by Claudius Ptolemy of Alexandria, who in the second century AD, produced a geography of the World in eight volumes. There were others and the dissemination of such books, even before the use of the printing press, was widespread. Writing was known in the Italian peninsula from at least the start of the eighth century BC, the date of the earliest examples so far found there. Latin inscriptions are known from the sixth century BC, with texts on law and ritual from shortly thereafter; the earliest surviving works of Latin literature date from the mid-third century BC, roughly the end of the First Punic War, and joined existing Etruscan and Greek works. Literacy became widespread. The acquisition of books was the mark of a cultured person and wealthy people built up collections, private libraries. By the mid-second century BC, 'there were rich library resources in the city of Rome.'[5] Map-making stretches back to the stone age and development of the art by Egyptians, Babylonians and Greeks brought cartography to a high level of sophistication and as far as the nearer areas around the Mediterranean were concerned, reasonable accuracy.[6]

An understanding of one's place in the world and of what surrounds it is essential when considering and planning for any expansion and, once the Roman state had grown beyond the mere parochial, it too had to consider the effect of any move by it against a much larger background. This was especially so in the maritime sphere, where large distances separated centres and the limitations in the operation of ancient shipping added new dimensions to Roman military thought. Consideration of the sheer scale and nature of Roman military operations demonstrates that their geographical knowledge was extensive and Roman expeditions were not simply sent off into the blue but followed detailed plans and maps. As an example, Poseidonius (who died in Rome in 51 BC) wrote a detailed description of Gaul giving a geographical overview of the country as a whole which must have proved invaluable to Caesar in his later conquests. The detail in the descriptions related by Poseidonius, a resident of Rome, in the absence of travelling the length and breadth of the country, came from perusing maps and accounts of the country of a good and complete kind. Similarly, when Augustus planned his advance of the Empire's borders to the Danube in the twenties BC, it was staged with a methodical advance, firstly to the line of the River Sava, then to the parallel Drava and finally to the Danube, again indicating detailed knowledge of the territory. Finally and on a truly strategic scale, in the third century AD, the emperor Gallienus (AD 253–68) withdrew from the Agri Decumates to avoid being out-flanked by the breakaway 'Gallic Empire' of Carausius, leaving a wedge of barbarian-occupied land between them. There survive other geographical accounts, all of which underscore the detail acquired by the Romans of their surroundings.[7]

From a purely naval perspective, the first strategic move was the Roman occupation and colonisation of the Pontine Islands in 312 BC. This group of islands has good anchorages and is situated about 25 miles (40 km) off the Italian coast, opposite the Gulf of Gaeta. The burden of having to maintain and supply a garrison on the archipelago was more than offset by its strategic position, halfway between the mouth of the Tiber and Ostia in the north and the ports of the Bay of Naples in the south and thus able to control and protect that vital sea route. Similar considerations can be seen in the establishment in the mid-first century AD, of the Roman Black Sea naval base at Chersonesus which they maintained for nearly 300 years. The position, only 100 miles (160 km) from the port of Sinope on the south coast and 120 miles (200 km) from the mouth of the Danube, was ideally placed to oversee and act as a listening post for the whole of the Black Sea area, as well

as a point from which power could be projected in any direction

The Punic Wars forced the need to develop strategic planning beyond the Tyrrhenian Sea. The area of operations of the First War concentrated on Sicily but to have been able to see the significance of Sicily and its islands, the Romans had a clear map-like view of the Tyrrhenian Sea and relative positions of Corsica, Sardinia, Sicily and the smaller offshore islands, as well as the Sicilian Narrows and North African coast. The relative merits of differing sea-routes and the relative positions of cities and harbours was also known. This can be illustrated by the battle of Ecnomus, which took place off the somewhat exposed, unfriendly and harbour-free southern coast of Sicily. Sailing with their invasion fleet for Africa, the Romans could have proceeded along the safe, sheltered north coast of Sicily, with its plentiful, friendly harbours, rounded the west of the island and crossed to Africa. To do this, they would have to have run the gauntlet of passing by the enemy strongholds of Trapani and Marsala at the western end of Sicily. Instead they chose the riskier southern route, knowing the enemy fleet was in the vicinity, but having a much shorter crossing direct to Africa and a clear line of communication behind them.

In the Second War, the area increased even more, taking in the whole western Mediterranean basin and extending eastward to western Greece. Once again, it was their knowledge of the geography that enabled Roman strategic planning to rate and apply themselves to the respective theatres of operations and to deploy their forces accordingly.

Finally in the second century BC, the whole of the Mediterranean Sea came under Roman domination. With the subsequent Imperial expansion, the whole of western Europe, the North African and Black Sea littorals, Asia Minor and the Middle East were absorbed. Missions of exploration were sent by Agricola around the British Isles and to the Orkney Islands and into the Baltic approaches.[8] Nero sent an expedition up the Nile which reached the Sudd area of southern Sudan and his general

Corbulo probed the Caucasus in the east. Such expeditions show that the Romans had an extremely clear appreciation of geography and the relative position and importance of places to enable planning on a strategic scale.

Deployment

This appreciation of the strategic position can be seen in the way in which the Romans deployed their fleets at various times, reflected in the organisation of the fleets and the level of operations in differing areas and to meet differing perceived strategic or political needs. As examples of this latter, in 204 BC, an embassy was sent to King Attalus of Pergamum on five quinqueremes 'as befitted the dignity of Rome';[9] during the Second Punic War, in 206 BC, Scipio sailed in a quinquereme to persuade the king of Numidia to become an ally.

Deployment for the First Punic War started in a reasonably simple way, with the main body of the fleet concentrated for the descent on Sicily, the focus of the war. The Roman's initial move against the Lipari Islands to the north of Sicily and their victory at the battle of Mylae (260 BC) gave them the strategic position to dominate the north Sicilian coast as well as the Straits of Messina between the island and the mainland. The archipelago could also guard the southern Tyrrhenian Sea and the approaches to Sardinia. Whereas guard ships were posted at various locations – Naples, for example – the fleet operated in the main as a single striking force throughout the war, the enemy operating in a similar manner.

The fleet itself was the instrument used to effect strategy in 229 BC. Troubles on the Dalmatian coast led to a Roman response whereby an army was transported across the Adriatic from Brindisi to the Albanian coast. A fleet of no less than 200 ships was also sent, which brushed aside the negligible naval opposition. The concentration and use of virtually the whole of the navy for this one operation might appear excessive but the demonstration of such overwhelming might was effective. The army with such backing was quickly able to subdue such opposition as did not surrender.

The result was the ceding of territory to Rome – roughly the area of modern Albania – which gave them control of both sides of the southern Adriatic, a reduction in the endemic piracy of the Dalmatian coast and a thwarting of the expansionist ambitions of Macedonia and further, gave it influence in Greece. All of this was achieved with little loss to the army and what amounted to a training exercise for the navy.[10]

With the Second Punic War and its far wider-ranging theatres of operations, deploying the fleet had to become more sophisticated in order to encompass all areas. Militarily, the war concentrated on containing Hannibal in Italy, campaigning in Spain, containing the Greeks and, ultimately, taking the war to the enemy homeland. It was for naval forces to dictate the direction of the war, assisted by the absence of determined Carthaginian naval opposition, and fleets had to be manipulated accordingly.

By considered deployment of their naval forces, the Romans were able to in turn manipulate the circumstances that enabled them to win. This is best illustrated by examining the way in which fleets were deployed during that war. At the start of the war in 218 BC, the Romans had a fleet strength of about 220 combat ships, requiring between 50,000 and 60,000 men. One consul was ordered with 160 ships to Sicily, with a view to menacing the Carthaginian North African homeland and another fleet of sixty ships was sent to cover the coasts of Gaul and Spain.[11]

More flexibility and a reaction to changing circumstances is apparent in Roman dispositions for 216 BC, when their plans called for the fleet of 220 ships to be divided with fifty-five ships to be sent to Spain and another forty-five to the Adriatic. This was backed by another fleet of seventy ships in home waters, able to reinforce the others as needed. Fifty more ships were laid up, to be a reserve of reinforcements or able to replace losses. The laying up of ships was necessitated by the need to divert manpower to the army, following the devastating losses caused by Hannibal's victories.

The commencement of hostilities in Greece

Early Roman warships followed the styles of the Etruscans to their north and the Greek settlers to their south. *Top*: An interpretation of an Etruscan 20-oared light warship, modelled after a fifth-century BC tomb painting at Tarquinia. *Below*: A Greek-style penteconter of the fifth century BC. Both these examples are monoreme, open craft, armed with a bronze ram and carrying a sailing rig, which can be lowered, with a single, rectangular sail.

With the spread of Roman-controlled coastline came the need to strengthen their presence at sea and to adopt more powerful warships. *Top*: a copy of a fourth-century BC terracotta model of a warship from Cyprus. It could represent a bireme with a decorative crenellated bulwark as it has no outrigger to carry the thranite oars and the oarports below are deliberately emphasised with coloured circles. Alternatively the indentations in the bulwark could be for oars and the model show an undecked trireme. *Below*: Author's model of a fifth-century BC Greek-type trireme, probably also in service with the Italiote Greek allies and similar to those that fought the Persian and Peloponnese Wars.

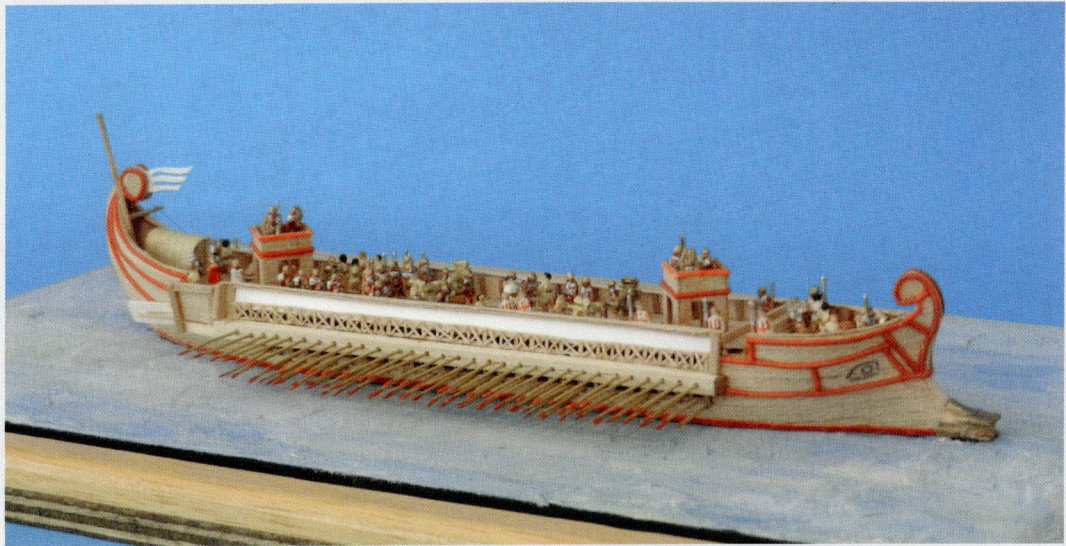

The mainstay of Roman battle fleets of the Punic, eastern and civil wars between the third and first centuries BC was the quinquereme. *Top*: Author's model of a later type, equipped with two towers, four artillery pieces and a light boarding bridge (which replaced the *corvus*). *Below*: a squadron of Roman warships at sea. The ship at front right clearly shows two remes of oars emerging from the hull side and a ventilation course between the oars and deck. The ships have single tower mounted amidships and the decks lined with marines. The ships are interpreted as quadriremes. Early first-century AD wall painting in the House of the Corinthian Atrium at Herculaneum. (*Author's photograph*)

Top: the *corvus*. Author's scale model of the device, according to the description given by Polybius, that gave the Romans their first victories over the Carthaginians. Shown mounted on the prow of a quinquereme, the stem post and forward bulwarks have been omitted to allow the *corvus* a full range of deployment.
Below: a small pleasure boat, typical of the less ornate scaphae or skiffs, carried by warships as tenders. Early first-century BC wall painting. Palazzo Massimo, Rome (*Author's photograph*)

Without enemy battle fleets to face, Roman warships were adapted to peacetime roles.
Top: warships entering and leaving harbour, first-century AD wall painting from Pompeii and thus probably giving an impression of ships at Misenum, just across the bay. Both ships have more than one reme but it is not clear whether they are biremes or triremes; neither has a tower so they are not any larger type. Both have marines on deck but the form of the sterns differs, perhaps indicating different types (*Author's photograph*).
Below: Author's suggested reconstruction of a light scouting craft of 20 oars, of the first centuries BC and AD. The model is purely conjectural and based upon wall paintings of light craft of that period.

For most of the imperial period, the Empire's eastern European borders rested on most of the length of the Rhine and Danube and the river fleets evolved their own types of warship.

Top: a modern reconstruction of a late first century AD small warship of 18 oars, based upon the remains of one of the examples found at Oberstimm on the upper Danube.

Below: a model of a fleet base on a tributary of one of the main rivers. Slipways and shipsheds are provided for the ships of the squadron based there, together with barracks (to the right) for the garrison. (*Koehler Verlagsgesellschaft mbH*)

By the second century AD, warships had ceased to have waterline armoured wales and protective decks over the rowers, a light lattice framework that could be covered by screens and awnings, giving sufficient protection to the rowers from enemies without artillery. The ships had smaller, upturned rams designed to ride over and swamp the light, open barbarian craft, supported by prominent fighting platforms forward.

Top: in this scene from a copy of Trajan's Column, two bireme liburnians can be seen, transporting troops for the Emperor's Dacian campaigns. The men are shown oversize and (erroneously) rowing through the side screens. Ahead is a military transport of a standardised type, from which stores are being unloaded (*Archaeological Museum, Bucharest. Author's photograph*).

Below: Author's model of the early second-century trireme shown on Trajan's Column in Rome as the emperor's personal transport.

Attempting reconstructions of late period Roman warships is made difficult as little evidence survives as to their appearance, save to show that their forms had changed almost completely from those of earlier periods. *Top*: Author's (incomplete) model of a *scapha exploratoria* of the British Fleet, late third century onward. The model was built shell-first on a mould and removed for insertion of the ribs, stringers and other internal timbers. The deck is being planked with fine strips, spaces being left for the hatches. Made from lime with pine decking; model length is 5.5 inches (140 mm), scale 1:192.

Below: Author's model a fifth-century AD warship, to a smaller scale (1:300). As little detail is known of these ships, only an overall impression can be attempted. Model length is 4.5 inches (114 mm). The ship is of modest size (110 feet/ 33.5 m overall), a monoreme type with the oars double-banked and a permanently fixed rig. The ram is an upturned spur, designed to override and swamp its opponents.

in 215 BC meant that a squadron of twenty-five ships had to be stationed in the Ionian Sea for operations in Illyria and Western Greece. By 208 BC, the navy had about 235 ships in total, manned by approximately 70,000 men. This force was deployed with twenty-five ships in Greece and another thirty at Taranto, intended to contain Hannibal but also from where they could be sent east to Greece or west to Sicily; a further seventy ships were in Sicily and fifty more in Sardinia, from where they could reinforce Sicily or Spain. In Spain, where by now there were no significant enemy naval forces, there were another thirty ships; finally, another thirty or so ships were in process of refit.[12] It might appear that this apparent dispersal of forces would leave the navy spread too widely and thinly and open to a concentrated, surprise attack by the Carthaginians. The Romans had, however, to cover their internal supply routes, as much of Italy's grain and food had to come from the islands of Sicily and Sardinia, to compensate for the loss of domestic production caused by Hannibal's presence in Italy. Ships were also stationed to attempt interception of supplies, reinforcements and despatches to Hannibal. Further, they had to maintain their communications and supply route to the Roman army in Spain and finally, to isolate and contain the affairs in Greece. The Carthaginian navy was itself well past its best and probably unable seriously to challenge the Roman fleets in Sicily, which in any case, could be augmented from the other fleets to oppose any such threat.

Dispositions changed again in 204 BC, with the final campaign in North Africa which ended the war. Forty ships accompanied Scipio's invasion fleet, with another forty in Sicily in close support; a further thirty were at Taranto and forty more in South Italy. Yet another forty ships were in Sardinia and thirty in Spain. Once more there were thirty or so ships in course of refit, indicating that there was a continuing, rolling programme of refitting ships.[13]

With the end of the Punic Wars and the eradication of any naval opposition in the western Mediterranean, the focus of naval activity shifted east. The following decades saw a Roman fleet averaging between fifty and sixty warships operating in Greek and Aegean waters with its allies. In the eastern campaigns, the Roman allies Rhodes and Pergamum, with their respective fleets, enabled the elimination of the Macedonian, then the Seleucid navies, while Rome committed a naval force of limited size. Sicily continued to be the linchpin of the navy, a fleet of fifty ships being maintained there, which as before, could be deployed in any direction. Additional squadrons were kept in the Tyrrhenian and southern Adriatic and a large number of ships, left over from the wars, were laid up but kept in good condition.

These basic dispositions continued and operations by the respective units can be traced through the second century BC.[14] With the ending of the third Punic War (146 BC) and of the wars in Spain (133 BC) the need for large formations in the western Mediterranean was much reduced, as were the fleets there. With the conclusion of the Mithridatic Wars in the east in 85 BC, the fleet in the east was withdrawn and disbanded. With no reason for the continued upkeep of such a large force, the navy was not only reduced but neglected, and the fleets deteriorated and reached a nadir with only smallish numbers of ships in home waters.

This decline changed in 67 BC when, under Pompeius for the war against the pirates, the fleet was built up anew, enlarged and deployed in thirteen squadrons of Roman and allied ships, each to cover a specific section of the Mediterranean, backed by a further mobile squadron of another sixty ships. At the end of that war and with no significant naval opposition left anywhere in the Mediterranean or Black Seas, the strategic imperative for the navy changed completely. Instead of forming fleets and squadrons for particular purposes and to oppose other fleets, it had in effect, become a police force and had to be formed into units suited to defined areas and duties. As such the navy was thenceforth deployed in formations, each with its own area of responsibility and with permanent bases. These areas increased with the progressive destruction or absorption of the navies of the various Hellenistic states of the east, completed by the acquisition of the Ptolemaic Egyptian navy in 30 BC.

The disruption of the civil wars of the forties and thirties BC led to these formations being concentrated in support of one or other of the antagonists and culminating in the two huge fleets of Octavian and Antonius at Actium. With the commencement of the Imperial age, once again the system of having a fleet unit for a defined area was re-imposed and became permanently formalised by the organisation of the fleets of the Imperial navy after 22 BC. Naval strategy thereafter, in the absence of any other naval opposition anywhere, became a matter of maintaining sea communications, the passage of trade free from piracy and the most efficient routes for the movement of armies and supplies: essentially the policing of the seas and rivers.

Later strategies

With the crises of the third century AD, and the deterioration of the Imperial navy, came increasing barbarian seaborne activity, presenting the Roman navy with its first naval opposition for nearly two and a half centuries. Initially the only practical strategy was one of reacting as best they could. The *Classis Pontica*, previously with its headquarters at Trabzon and positioned to dominate the south-eastern Black Sea, was moved in AD 152 to new headquarters at Cyzicus on the Sea of Marmora, initially to support opposition to Parthia in the Levant. The move became permanent later as, although still sending ships along the north Anatolian coast, the main purpose became to protect the Bosporus and entrance to the Aegean.

A clearer strategy emerged once the initial shock of a challenge at sea had subsided. The Romans retained their technological superiority and throughout the fourth century AD practised a more aggressive policy of interdiction against the barbarians. The difficulty was that unlike their opponents of yore, the barbarians did not have the cities and highly developed industrial and agricultural infrastructure that could be targeted. Enemies had to be located, destroyed

at sea or followed home so that their villages and fields could be attacked. The Roman naval policy had to be largely defensive, with the building on the coasts of Britannia and northern Gaul of system of forts (the Shore Forts), watchtowers and lookout points. The possibility of finding and intercepting raiders on the open seas was remote, especially with the limited ability of ancient warships to 'keep the sea'. Once enemy ships came within sight of the lookouts however, forces could be directed to deal with them. Although requiring a continuously manned string of lookouts and naval units to be posted at various points around the coasts at the shore forts, the strategy was on the whole successful.

Roman naval superiority in the Mediterranean was, despite some spectacular barbarian raids, not challenged until the Vandals occupied part of Spain and acquired ships. This was despite and imperial edict of AD 419 forbidding anyone from teaching them how to build or operate ships. The Vandals did get ships, crossed to Africa and eventually set up a kingdom at Carthage and by mid-century had seized naval superiority in the central Mediterranean.

On the Rhine and Danube, the great riverine frontiers, the strategy was first to prevent unauthorised crossings so far as possible; second, to control the barbarian banks of the rivers with military patrols, a strip of territory half a day's ride for cavalry, approximately 8–10 miles in width (12–16 km) could thus be maintained as a cordon sanitaire. Additionally, incursions were made up the tributaries on the barbarian sides of the great rivers, deep into alien territory.[15] This strategy remained effective until the deterioration of frontier forces, particularly from the fourth century AD, with a similar corrosive effect on the river fleets, allied to the increasing loss of control of the river frontiers ended it. The Danube was recovered by the end of the fourth century and naval forces on the river rebuilt by the emperor Theodosius (reigned AD 379–95). The former strategy was put back into operation and continued well after the fall of the Empire in the West.[16]

Intelligence gathering

The formation of strategy and the deployment of naval forces, together with the adoption of appropriate tactics for a prudent commander, depend upon the obtaining of intelligence about one's enemies. The gathering of such intelligence by spies, agents or scouts ashore (known as *speculatores*[1]) interrogating prisoners or reconnaissance by scout ships, were all methods open to and used by the Romans. Time and again, ancient fleets were despatched to intercept or counter enemies, with fore-knowledge enabling them to head in the right direction to find them, rather than simply wandering around in the vain hope of an encounter.

Good gathering and evaluation of intelligence proved at times decisive such as in 241 BC, when the Roman fleet commander, Lutatius, learned of the approach of a heavily laden and undermanned Carthaginian relief fleet from Africa, bound for their garrisons on western Sicily. With this knowledge, the Roman fleet was able to position itself to intercept the enemy and destroy it at the battle of the Aegades Islands, which ended the First Punic War.[2] An equally impressive intelligence coup occurred in 215 BC, early in the Second Punic War, when the Carthaginians sought to secure Sardinia, lost by them to Rome after the first war. They sent sixty warships with a military force to foment rebellion and raise the island, a major provider of food for Rome, against her. The Romans had discovered these plans and sent a fleet and a legion which landed ahead of the Punic force and defeated it when it landed.[3] In both of these instances, intelligence of enemy plans and moves had to be gained from sources in enemy territory, in Africa and communicated to the Roman commanders in time. This in turn requires a network of agents or spies, but also a system for the transporting of reports overseas.

Such a system would have to be on demand with locally available fast craft, rather than by waiting for prearranged clandestine rendezvous with Roman scout ships. An incident, perhaps apocryphal, demonstrates the speed of travel of such reports, namely that in about 150 BC, in inciting the Third Punic War and to emphasise the dangerous proximity of Carthage, the Senator M. Porcius Cato produced in the Senate House a ripe fig which he claimed had been picked in Carthage only three days before; the Senators knew such a voyage to be possible.[4] The gathering of intelligence by covert means was augmented by the use of scouting or 'spy' ships, working inshore to patrol designated coastlines. This practice, pursued by the Romans in the First Punic War, enabled them not only to gather information about the enemy, but, backed by large-scale naval raids, to foment insurrection by subject peoples of the Carthaginians in North Africa, requiring them to direct a considerable part of their military effort there to contain the revolts, to the extent that they were left with insufficient manpower to crew all but a few of their warships.

The value of scouting ship patrols was also proven in 215 BC, when enemy ships were spotted by them off the Calabrian coast. A force was sent and captured what proved to be envoys with the whole documentation of a treaty between Hannibal, then rampaging around Italy, and King Philip of Macedon, who was to invade Italy and join him. With them both in Italy, they could have gone on the offensive in a strength that could well have proved fatal for Rome. Philip, not knowing if he had concluded a treaty and thus an ally to meet him, never came and the Romans were able to take steps to keep him committed in Greece and avoid the union of their enemies.[5] Scouting of shores was not always so successful and in 55 BC, prior to his first landing in Britain, Caesar sent

23 THE PERIPLOUS BATTLE

Mylae 260 BC

The Roman fleet sailed from its base at Messina, rounded Cape Pelorus and headed westward along the north coast of Sicily. The fleet comprised 143 ships, mostly quinqeremes, with some triremes; some of the latter, together with smaller, pentecontars ahead of the battle squadrons as a scouting screen. The Romans had heard that the Carthaginian fleet was active in the area of Milazzo (Mylae), and were keen to bring them to battle and secure their flank on the north coast of Sicily, also to try their newly enlarged fleet. The Romans were in battle array of line abreast and as they approached, they were spotted by Punic lookouts. The Punic fleet

had a strength of about 130 ships, again, mostly quinqueremes and with some triremes and was commanded by Hannibal, flying his flag in a 'seven'.[1]

Supremely confident in their ability and not a little contemptuous of the mostly inexperienced Roman fleet, the Punic fleet emerged from its anchorage at Milazzo. As they formed, they sent about thirty of their fastest ships ahead to engage and stop the centre of the Roman line, hoping that the momentum of this would cause the outer parts of the Roman line to coalesce to some degree on the centre. The remainder of the Punic fleet, in two parts, intended to then go around the flanks and rear of the Roman line

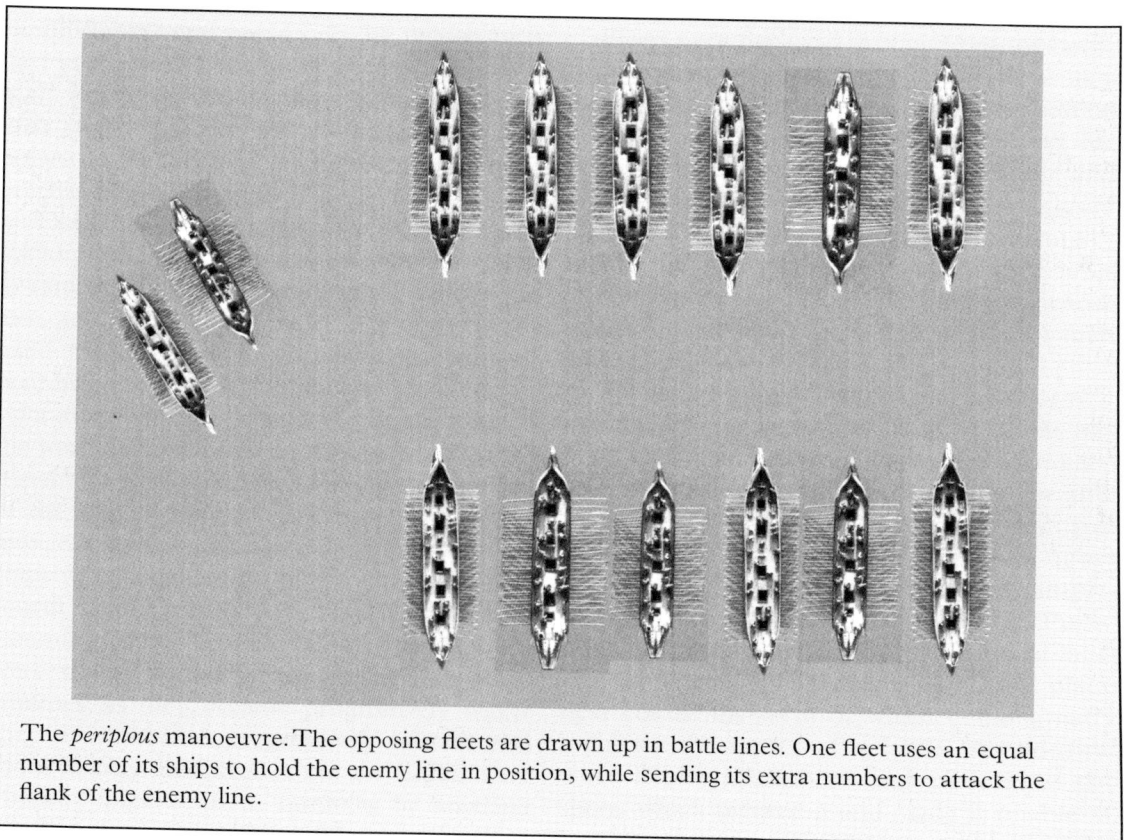

The *periplous* manoeuvre. The opposing fleets are drawn up in battle lines. One fleet uses an equal number of its ships to hold the enemy line in position, while sending its extra numbers to attack the flank of the enemy line.

118

to execute a double *periplous,* ramming their enemy in the beam and after quarters.

Fully aware of their comparative lack of experience in operating large fleets and in ship handling, as well as of the ability of their opponents, the Romans had taken measures to lessen those disadvantages. As it was, their quinqueremes were heavier and slower and less handy than those of the Carthaginians, so they erected towers on the decks, each capable of holding up to six archers, the first time that these had been used on a seagoing ship for fleet actions.[2] The normal complement of marines on a quinquereme was forty men; to augment this, a century of soldiers from the legions was put aboard each ship. The Romans' best weapon was their infantry, which was now afloat in overwhelming force.

Finally, to enable these men to board their enemy, the ships were fitted with the *corvus.* These must have been prefabricated and taken down to Messina by transport ships, as they were not aboard the ships of the fleet when they made their way down the coast to Sicily. Two previous brief encounters with the Punic fleet had failed to reveal them to the enemy and Polybius reports[3] that the Carthaginians were puzzled at the sight of these contraptions as they sailed out to battle at Milazzo. The Carthaginians were very soon to learn to their cost, the meaning of those devices when their lead squadron engaged the Roman centre, whereupon they were impaled by the raven's beaks as the *corvus* was dropped to their decks, followed by the mass of infantry and marines that quickly overcame them. All thirty of the ships soon succumbed, including the flagship of Hannibal, who escaped in the ship's boat.

With this success and impetus, the unengaged Roman flank squadrons did not obligingly concentrate around their centre, but turned outward, to meet the oncoming main body of the enemy fleet. These formations saw what was happening to their lead ships and that their attempt at *periplous* was about to be thwarted and tried to sheer off and avoid the oncoming Romans. Not all were successful and the Romans swung their *corvus* to whatever beam

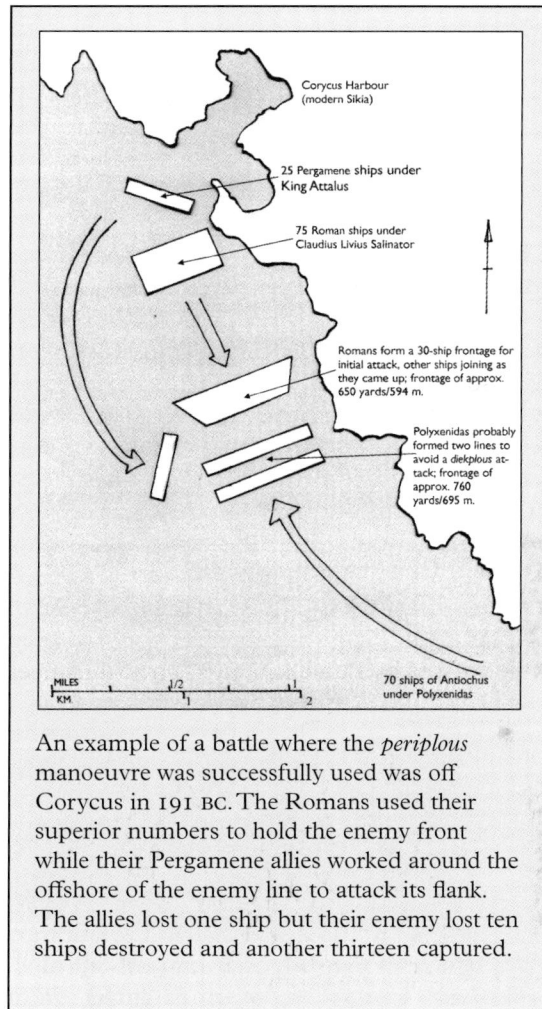

An example of a battle where the *periplous* manoeuvre was successfully used was off Corycus in 191 BC. The Romans used their superior numbers to hold the enemy front while their Pergamene allies worked around the offshore of the enemy line to attack its flank. The allies lost one ship but their enemy lost ten ships destroyed and another thirteen captured.

was nearest and managed the ensnare another twenty Punic ships. By the end of the battle, the Romans had sunk or captured fifty Punic ships, including the flagship, inflicting some 7,000 casualties and taking 3,000 prisoners, for the loss of only one ship and a few hundred casualties.

Although the larger part of the Punic fleet escaped, the battle proved to be decisive as the Carthaginian's long assumption of their superiority at sea, along with their dominance of the western Mediterranean had been shattered and with it, their confidence, never to be regained. Rome on the other hand, had become a world-class naval power.

Carthaginian tactics had been sound in

seeking to use moves that would capitalise on the superior handling and speed of their ships; the fact that so many of their ships did manage to disengage or evade and to escape is evidence of that ability. Had the Punic centre not been so quickly overwhelmed, the Roman line, even by keeping station, would have been slowed almost to a stop, leaving them highly vulnerable to the sweeping *periplous* manoeuvres of the second line of Punic ships, around their flanks and rear. As it was, the Roman flanks were able to turn to meet and thereby thwart those intentions, implying an independence of command and displaying perhaps, better abilities and initiative than was expected from them and from their initial tightly maintained formation.

Naulochus 36 BC

The final battle of the civil war, between Octavian and the younger son of Pompeius the Great, Sextus Pompeius, was in late August or early September off Naulochus in northern Sicily.[4] It is a good example of the successful use of the *periplous* manoeuvre. Sextus' fleet was of about 120 ships, including one six, a few quinqueremes and some quadriremes, but otherwise, smaller types; they were commanded by Sextus' admiral, Demochares, a freed slave. Octavian's fleet was commanded by his close friend, Agrippa and was about ninety-five strong but included many sixes and quinqueremes and which were also equipped with the *harpax*. Both sides' larger ships mounted towers and those of Agrippa were all painted in a uniform colour for easy identification.

The two fleets approached each other arrayed in line abreast, but Demochares, mindful of the smaller average size of his ships, deployed one of his flanks close inshore, so that it could not be turned and held the rest of his line in a very tight formation. Agrippa conversely, allowed his ships a looser formation, taking advantage of his opponent's comparative lack of space to manoeuvre. Securing the front of the line with a general engagement, Agrippa extended his line to seaward, enveloping Sextus' offshore flank in a textbook *periplous* move. Agrippa's bigger and heavier ships used the *harpax* to ensnare

A gold four-denarius medallion of Augustus. He is entitled *pater patriae*, 'Father of the Country', a title bestowed upon him by the Senate in 2 BC. On the reverse is figure of the goddess Diana and the letters SICIL, indicating that it was issued to commemorate the victory over Sextus Pompeius at the battle of Naulochus in 36 BC. Naples Archaeological Museum. (*Author's photograph*)

their opponents, forcing them back upon their fellows, at the same time, denying them any searoom in which to manoeuvre or escape and rolling up their line towards the shore. Twenty-eight of Sextus' ships were sunk, only seventeen managing to escape and the rest being forced on to the shore and destroyed or burnt. The escaping ships dumped their towers overboard to lighten them and, with Sextus on board, fled to Asia, where he was later executed by Antonius. Agrippa lost only three ships.

By not taking advantage of the better speed and mobility of his lighter ships, Demochares had surrendered the tactical initiative to Agrippa and indeed limited his ships to the worst possible situation, where they could make no response to counter the weight of Agrippa's ships. Agrippa clearly learned this lesson as, when the position was reversed at Actium, it was he who had the larger number of lighter ships against Antonius' preponderance of larger, heavier ships.

24 THE DIEKPLOUS BATTLE

Myonnesus 190 BC

There was intense naval activity in the eastern Aegean, along what is now the western coast of Turkey, during the war between Rome and her allies, Rhodes and Pergamum and Antiochus, the Seleucid king, between 192 and 189 BC. The final naval battle of that war took place just north of Cape Myonnesus in September or October of 190 BC, near the end of that year's sailing season.[1] The Seleucid fleet, under the command of its admiral, Polyxenidas, comprised eighty-nine ships, including two sevens and three sixes and manned by approximately 24,000 men. The Roman fleet, commanded by Aemilius Regillus, was of fifty-eight ships, mostly quinqueremes, with an allied squadron from Rhodes, commanded by Eudamos, of twenty-two lighter ships, quadriremes and trihemiolias, a type much favoured by the Rhodians and totalling some 20,000 men.

Polyxenidas put to sea from his base at Ephesus and sailed the short distance to anchor at the small island of Macris where he was concealed by the cape. His intention was to sprint the short distance, only about 10 miles (16 km), in the hope of surprising the Roman fleet at their harbour of Teos. This harbour had only a narrow entrance and would act as a trap. The Romans however, received intelligence that enemy ships were laying near to the cape and

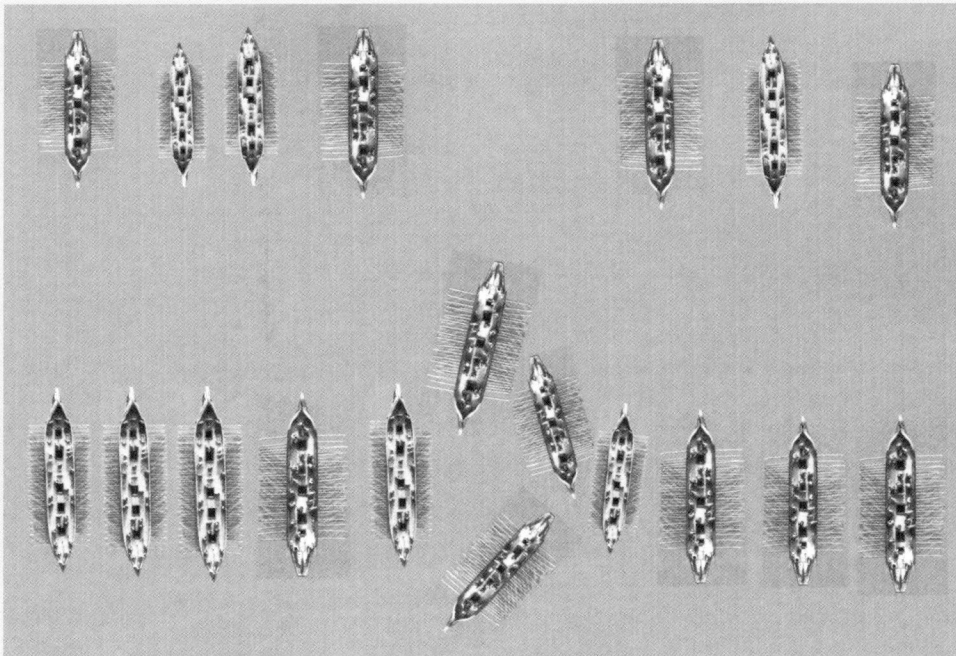

The *diekplous* manoeuvre. A determined attack is made to force a gap between ships in the enemy line. The leading attacker will be at least severely damaged but if successful, allows the ships following it to break through the line and to attack its rear.

they therefore managed to put to sea albeit in considerable haste. The fleets met just north of Cape Myonnesus.

Polyxenidas deployed his ships into two lines abreast, one behind the other, with his left flank close inshore and extended his offshore flank to seaward, hoping that his superior numbers would envelop the Roman right or offshore flank. For their part, the Roman fleet was arrayed in line abreast, with the Rhodians behind them as they had been the last to leave the harbour. As the opposing fleets neared each other, Eudamos led his ships out to seaward, emerging beyond the Roman line, to cover their flank. There, the Rhodian ships harried and held the attempted Seleucid *periplous* manoeuvre. In the meantime, the Roman ships, with little or no space for manoeuvre, had little option but to head straight on to engage their opposite numbers. The Romans pressed on until the right (seaward) end of their line started to force it's opponents back and then they broke the Seleucid line in a classic *diekplous* attack, emerging in the rear of the enemy ships. Using their well-honed grapple and board tactics, the Roman marines fought methodically, overcoming one enemy ship, then proceeding to the next in turn.

The Rhodians used their swift, handy ships to attack with the ram and missiles, while avoiding becoming entangled by the larger enemy ships. They also deployed fire baskets, slung from long spars projecting from their bows,[2] which deterred enemy ships from closing and in forcing them to turn away to evade, exposed their sides to ramming.

Under the Rhodian attack, Polyxenidas' offshore wing started, before long, to be forced back; his centre meanwhile, was being systematically destroyed. The offshore wing started to retreat and the other survivors of the Seleucid fleet soon followed, running for the safety of their harbour at Ephesus. Polyxenidas had lost thirteen ships sunk or burned and another thirty-nine captured, nearly half of his fleet. Roman losses were two ships and several more damaged, the Rhodians lost one ship.

The battle was essentially of two parts; alone, the Romans may have been enveloped and eventually worn down by the sheer weight of the enemy; the Rhodians for their part, could not venture alone against Polyxenidas. Eudamos' ships, with their agility and the skill of their crews and handling was clearly the critical factor in that by disrupting the Seleucid *periplous* attempt, the effect of their superior numbers was largely negated and the Romans could concentrate on doing what they did best. The successful *diekplous* attack by the Romans was the *coup de grâce* which led to the large enemy losses.

Actium 31 BC

The naval battle of Actium was the decisive last battle of the civil wars of the first century BC, which ended them and left the victor, Octavian as sole ruler of the Roman World and first emperor. It was also perhaps, the greatest missed opportunity for a perfect *diekplous* attack.

The small harbour of Actium lies on the southern shore of the narrow entrance to the Gulf of Ambracia, in western Greece. Antonius and Cleopatra were encamped with their forces near the harbour, with their fleet safe within the Gulf. They had fortified positions either side of the 400-yard-wide (365 m) entrance to the Gulf, to block any attempt by their opponents to enter. Octavian, with his army was encamped to the north of the entrance, with his fleet, again commanded by Agrippa, drawn up and moored by the adjacent foreshore.

By late August Antonius' forces, although in a very strong defensive position, were in an increasingly parlous state, becoming very short of supplies and suffering desertions. Whereas Octavian refused to be drawn into a conventional battle, where Antonius, a very experienced general, would have a numerical advantage, his cavalry sweeps were cutting Antonius' overland routes of supply. With the enemy fleet safe but inactive within the Gulf, Agrippa had used his fleet to completely cut Antonius' lines of supply by sea, the method upon which he had in fact principally relied. In an increasingly desperate situation and without a secure land route, he and Cleopatra had little option but to consider a break-out by their

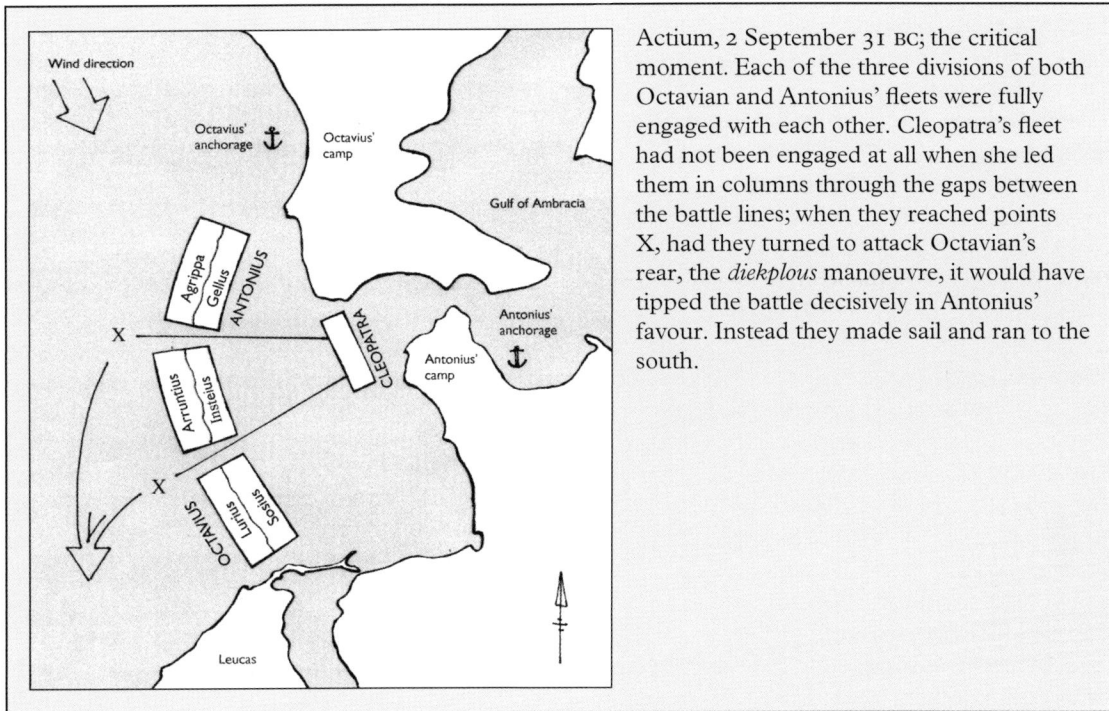

Actium, 2 September 31 BC; the critical moment. Each of the three divisions of both Octavian and Antonius' fleets were fully engaged with each other. Cleopatra's fleet had not been engaged at all when she led them in columns through the gaps between the battle lines; when they reached points X, had they turned to attack Octavian's rear, the *diekplous* manoeuvre, it would have tipped the battle decisively in Antonius' favour. Instead they made sail and ran to the south.

combined fleets, to force conclusions in a sea battle. It had to be a mass breakout as an earlier attempt by one of his squadrons to break the blockade had been beaten back.

This was in fact for them, a reasonably advantageous course as between them they commanded fleets, on the face of it, the equal, if not superior to that of Octavian and Agrippa. Antonius had just over 120 heavy warships, including 'tens' and 'sevens', the last of the Hellenistic 'super-galleys' or polyremes that had been evolved and built during the naval arms race between the rival successors of Alexander the Great. These ships had been gathered from where they had been laid up in the Levant after dissolution of the navies, such as the Seleucid and refitted. In addition to the towers and artillery with which these ships were equipped, there are references to raised decks or platforms being erected, to give their marines an additional height advantage. They were described as 'being mounted with towers and high decks, they moved along like castles and cities while the sea groaned and the winds were fatigued'.[3] In addition, there was Cleopatra's Egyptian fleet,

of about sixty ships, including more of the polyremes of which the Ptolemies had been so fond. This battle was in fact to be the last foray of the Hellenistic polyremes and the only time that they formed part of a Roman fleet.[4]

They loaded these ships with some 20,000 troops and 2,000 archers, in addition to their normal complements, just as many of Antonius' men as could be got aboard. They sought to overwhelm Octavian's ships by sheer weight of numbers. If they achieved their break-out and beat Octavian, the men who had to be left behind would have been relieved, if not, they were abandoned, with obvious consequences for morale and loyalty; as many as possible had to be seen to be taken.

On 2 September the fleet formed in the entrance to the gulf into four squadrons, of about forty ships each, in several lines, emerging through the entrance to deploy into battle order. One squadron under Gellius Publicola, with Antonius aboard, formed the right flank; the second squadron, under Marcus Insteius and Marcus Octavian (no relation) was the centre of the line; the third squadron under Gaius Sosius

Actium, the area of the battle viewed from the site of the ancient Greek city of Kassope. The Gulf of Ambracia is to the left, Antonius and Cleopatra's fleet anchorage. The entrance to the gulf can be made out; Antonius and Cleopatra's camp was on the far side, that of Octavian on the nearer, with his fleet anchorage in the bay on the right. The island of Levkas is in the distance on the right and the battle took place off the gulf entrance between the bay and the island. (*Author's photograph*)

taking the left flank, a total of nearly 170 ships. Cleopatra with her Egyptian fleet of sixty ships took station in the centre, behind Insteius.

Alerted to the great activity in the enemy camp, Octavian's own fleet put to sea.[5] The fleet had a little over 200 ships, with a few sixes, quinqueremes, quadriremes and triremes and a lot of the lighter types of ship, including those captured from Sextus Pompeius. The fleet was largely that which had won at Naulochus five years earlier and was thus battle-hardened and with its more recent operations, at the peak of morale and condition. Antonius' fleet was by comparison, inexperienced and untried in battle.

By about noon, the two fleets were in their respective battle formations, about 1,600 yards (1,463 m) apart. Octavian's fleet deployed into three divisions, each of about sixty-five ships; his left was commanded by Agrippa, Lucius Arruntius commanded the centre division with Octavian himself aboard, with the southern (right) division commanded by Marcus Lurius. Antonius' ships formed an arc of about 500 yards (457 m.) between the northern and southern banks of the entrance to the gulf, the flanks secured by the banks, while Octavian's fleet formed a matching arc to seaward of about 800 yards (731 m.). Antonius, in the stronger defensive position, hoped to entice his enemy to

action and thereby break their tight formation, but Octavian refused to be drawn, so Antonius had no option but to attack.

Gellius, on the right, was ordered to attack Agrippa and seek to turn his flank (*periplous*). Agrippa's ships had, meanwhile, been backing water and widening the distance and thus drawing Gellius further forward, out of line. Sosius, in the south next advanced, but as they got further from the gulf into more open water, Antonius' formations became extended and began to lose cohesion. At this point, Octavian signalled a general engagement.

Agrippa had devised special tactics for the expected battle. The infantry on both side was of course, Roman and with his preponderance of big ships, Antonius sought a boarding battle in which Agrippa's smaller, lighter ships would be bettered. The big ships were like floating castles, with artillery and hordes of marines and virtually impervious to ramming by the light ships. They were however, slow in speed and manoeuvrability and so several of Agrippa's lighter ships were to single out and attack the oars and rudders of a large ship, avoiding its grapples and ram but themselves ramming as opportunity arose; although having little individual impact, the cumulative effect of several hits in the same locality could and did cause damage. This was facilitated by the lines becoming more extended as they advanced to open sea and trying to avoid a *periplous* attack. Additionally, they shot fire arrows and pots of burning pitch and charcoal from their catapults.[6]

After an hour or so of intense fighting, Antonius' centre (Insteius) and left (Sosius) began to fall back but the battle was still very far from over or decided. Cleopatra then formed her division, which had not yet been engaged, into columns and proceeded to seaward, passing between the divisions of the fleets which were locked in combat, into open water behind them. This was the perfect *diekplous* and Octavian and his admirals must have been only too aware that, with no reserves left to oppose the Egyptians, a defining moment had come. Had Cleopatra's ships then and presumably as expected, fallen upon the rear of Octavian' squadrons, they would have been annihilated and the battle decided. With all of his ships engaged and despite their looser formations, an Egyptian attack from seaward into their rear would trap them and push them on to Antonius' floating castles, with no room to avoid the grapples and boarding.

As is well known, however, Cleopatra's ships hoisted sail and ran to the south, Antonius saw this and managing to extricate some fifty of his ships, followed her. The rest of his fleet, as they learned of their abandonment, surrendered or returned to harbour, to give up later. This was to be the last great naval battle in the Mediterranean for some 355 years and the last appearance in battle of the great Hellenistic polyremes.

Sulci 258 BC

By 258 BC Corsica and most of Sardinia, both formerly Carthaginian possessions, had fallen to the Romans. A Roman fleet, commanded by Caius Sulpicius Paterculus was operating in southern Sardinia, where a Punic fleet was also stationed. Each fleet was of approximately a hundred ships in strength, although their respective composition is not known.

The Punic fleet was moored or beached in the Gulf of Sulci in south-west Sardinia and despite the heavy sea mist, their lookouts spotted the Roman fleet sailing by the entrance to the gulf, apparently unaware of the presence of the Punic fleet.[1] The latter immediately put to sea in the belief that they were about to take their enemy completely by surprise. The Romans had however, been aware all along of the Punic presence and as soon as their ships came out to sea, the Romans turned and attacked them, in turn surprising the Carthaginians.

After a brief fight and after the Punic fleet suffered a few losses, the fleets disengaged and the Romans sailed off into the continuing mist whence they had come, until lost from sight. The Punic fleet returned to its anchorage and stood down, the crews going ashore for rest. Suddenly, using the mist and nearby high ground as cover, the Roman fleet reappeared and attacked the moored and virtually unmanned ships, inflicting further losses. The Carthaginians lost about forty ships in these encounters and their naval power in Sardinia and indeed, in the Tyrrhenian Sea was broken.

The Romans must have had their own intelligence of enemy dispositions and made good use of the nearby headland to cover their moves, as well as the fortuitous sea mist. This latter made the more plausible their initial, apparently unknowing sail-by that drew their

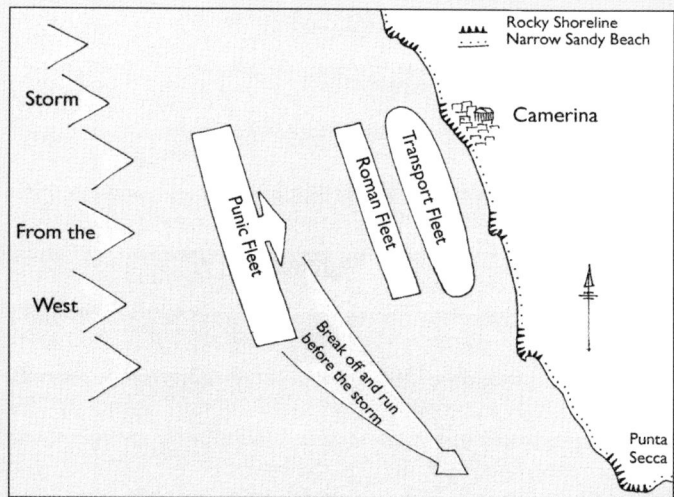

An example of a battle where the local topography played a defining role took place at Camerina, Sicily in 249 BC. The consul Lucius Iunius Pullus was retiring from Phintias with a fleet of some ninety warships, escorting several hundred transports, eastward to Syracuse. Caught by a Punic fleet of 120 warships under Carthalo, he was forced back and took up a defensive position to protect his transports, with his back to the shore. Forcing the Romans back further, before they became heavily engaged, the Punic fleet suddenly broke off the engagement and ran to the south-east. The westerly storm that the more experienced Punic commander had seen caught Iunius' fleet backed close into the shore, with its rocks and narrow beaches. The fleet was swept ashore and wrecked, only two warships and few transports escaping; the weather and shore had given the Carthaginians a victory.

The battle of Ecnomus, 256 BC. The town of Licata, the site of ancient Phintias, seen from the east, with the height of Ecnomus behind it. The long offshore breakwater is more recent. The Roman Fleet approached from the left, parallel with the shore; spotted by Punic lookouts on the heights, they were confronted by the Punic Fleet emerging in battle order from behind the headland. (*Author's photograph*)

enemy into a trap. Thereafter, once more, when hiding behind the headland, land-based lookouts can be presumed to have been posted to signal to the fleet when they saw the Punic ships moor and their crews go ashore, the proximity and cover offered by the topography making the surprise attack that followed, possible.

Ecnomus 256 BC

The First Punic War in Sicily had stagnated by 257 BC, the Romans holding eastern and central Sicily, but unable to dislodge the Carthaginians from their strongholds in the west of the island. To break the deadlock, the Romans resolved to make a landing in Africa with an army, to march on Carthage itself.[2]

To this end, the Romans built up their forces and concentrated their fleet at their bases of Messina and Syracuse. The fleet included two sixes as flagships, with the rest mostly quinqueremes, equipped with the *corvus*, with some triremes and twenty more triremes stripped and rigged as horse transports. There was also a convoy of eighty transport ships, for the invasion troops and their supplies. The Carthaginians had not been idle and had built up their own fleet in Sicily to just over 200 ships, which were operating from their forward base at Eraclea on the south-west coast, in anticipation of the Roman moves. They knew that the Romans would make their way westward along the mainly friendly (to them) south coast of Sicily, to embark their army at Agrigento.

Aware of the presence of an enemy fleet, although not of its position, the Roman fleet sailed in a battle-ready formation. Two squadrons of warships led, formed into a 'V' or wedge at the centre where they abutted with their outer flanks refused; behind them was a third squadron in line abreast. The transports were behind the third squadron and behind them was a fourth squadron of warships, again in line abreast, sandwiching the transports and acting as a rearguard.

The Punic fleet advanced to Licata (Ecnomus), where a headland juts into the sea (Mount Ecnomus), sheltering the small estuarine town of Phintias (modern Licata) to the east and offering a clear view eastward along the low-lying coastline of the Gulf of Gela. The Romans had to approach from this direction, their speed slowed to that of the transports. The Punic fleet was anchored on the west side of Mount Ecnomus and thus hidden from the Roman's view and they had posted lookouts on top of it. When the Roman fleet was seen approaching, the Punic fleet made ready and formed into battle squadrons in line ahead while still behind the headland, timing their advance for the last moment for maximum surprise. They emerged in column, each squadron turning to port in turn to form a line abreast across the Roman front and allowing the Romans no time in which to react or change formation.

Unfortunately for the Carthaginians and their carefully laid and executed trap, the Roman formations were the equal to it and their two leading squadrons, already in battle order, crashed into the two centre Punic squadrons, forcing them back. The unengaged Punic flank squadrons went around each side of the developing mêlée, to attack the other Roman ships. The two rearmost Roman squadrons turned to meet the enemy, the transports running for the haven of Phintias.

The rearmost Roman squadrons had to deploy from ahead and astern of the transports and were not fully ready when the Punic flank squadrons engaged and they were thus gradually pushed back against the shoreline. Seeing this and having driven off and scattered their opposite numbers, the two lead Roman squadrons broke off and rowed hard to attack the other two Punic squadrons in the rear. One of these saw the approaching danger and managed to disengage and largely to escape, the other however, was enveloped and annihilated. The Romans had lost twenty-four ships, all to ramming, but had sunk thirty and captured another sixty-four Carthaginian ships, using ramming, the *corvus* and boarding.

It is clear that had the Romans not already been sailing in battle order, the surprise appearance of the Punic fleet, already in battle formation, right across their front, would have given the Romans no time in which to deploy when they sighted the enemy. The Romans would have been caught unformed, in open order and the result of the battle might well have been very different.

Drepanum 249 BC

For a final example of the way in which topography could influence an ancient sea battle, Drepanum (Trapani) in 249 BC was unusual in being a defeat for the Romans.[3] By this year, the Carthaginian possessions in Sicily were reduced to the fortified cities and harbours of Marsala and Trapani in the far west of the island, the two being only 15 miles apart (25 km). Both were besieged by the Romans but could be and were supplied by sea. A Punic fleet of about a hundred ships under Adherbal, one of their best admirals, was at Trapani and the Roman fleet of 123 ships under the Consul Publius Claudius Pulcher was just south of Marsala. The Romans

learned that a second Punic fleet of seventy ships under another capable admiral, Carthalo, was en route to join Adherbal.

Pulcher had two options, retire eastward and join his fellow Consul, Lucius Iunius Pullus, who had a fleet of 120 ships, or to attack Adherbal before he could rendezvous with Carthalo. In either case, the Romans would outnumber their opponents, but the joint fleets of the Consuls would have a far greater superiority. Pulcher however, was hot-headed, impulsive and jealous to gain all of the glory for himself and decided to attack the enemy at Trapani, relying on his smaller advantage in numbers. He was regrettably a poor admiral and his fleet, having received many new drafts of men, was inexperienced and not fully worked up. The priests warned Pulcher that the sacred chickens would not eat and thus the auspices were bad, whereupon the commander threw the chickens overboard into the sea, commenting that if they would not eat, they could drink.[4]

The Roman fleet sailed at night, hoping to surprise their enemy at dawn, in port. They sailed in line ahead, northward along the coast, Pulcher being with his flagship at the rear of the column to prevent stragglers; this unfortunately meant that he was in the wrong position to command and could not see the circumstances that his fleet was sailing into, nor therefore, formulate or give orders to it. He also failed to consider the geography of the area in that, had he attacked from any other direction, he would have completely trapped his enemy.

As the Roman lead ships, still in line, made to attack, the Carthaginians saw them at the last moment and manned their ships in considerable panic and got to sea away from the Roman advance, by manoeuvring through channels between small islets just off the harbour shore, which screened them from the Romans (who should have known and anticipated this). This brought the Punic ships out to seaward of the Roman line. As they made the open sea, the Punic ships turned to port and fell upon the Roman ships which were thus broadside on to their attackers and still in an extended line parallel to the shore. Pulcher added to the

The battle of Drepanum, 249 BC. Trapani (Drepanum) viewed from the heights of Erice (Eryx). The Egadi islands can be seen offshore. The Roman fleet approached from the left, parallel to the shore, where the salt flats can be seen. The shape of the harbour has changed since the battle, but the islet which shielded the Punic sally can be made out. (*Author's photograph*)

confusion by trying to order a withdrawal, but the Roman ships, with no room to turn, were forced on to the shore and wrecked. The shore is flat and low and many of the Roman crews managed to get ashore and back to their own lines. It was a disaster, Pulcher, from the back of the line, managing to escape with thirty ships, the other ninety-three being lost and giving the Romans their only naval defeat of the war.

Had Pulcher attacked from any other angle, or even if their line had been advancing parallel to the shore but further out to sea, the Punic fleet would have been trapped in harbour. Any attempt by the Carthaginian ships to emerge from between the islets would result in their being picked off one by one as they emerged, any trying to escape southward could have been trapped against the shore as the Romans had been. The situation of the harbour, the coastline and above all, the existence and location of the islets had been used to full advantage by the Punic ships, taking full advantage of Pulcher's culpable negligence.

Carthage 203 BC

Towards the end of the Second Punic War the Romans invaded Africa, landing and fortifying a defensible promontory between Carthage, to the east and Utica, to the west. The invasion fleet of nearly 400 transports was attended by a close escort of forty warships, with another supporting fleet in southern Sicily, covering the line of supply between the two,

The Roman commander Scipio loaded his warships with artillery and siege equipment, intending to make a feint as if to attack Utica from seaward. Thinking that the Romans had gone off to attack Utica, the Carthaginian army came out to attack them in the rear and was promptly surprised and defeated and driven back into Carthage. Having thus gained the ground, Scipio advanced to the site of modern Tunis, drawing a line around Carthage on land.

The warships returned to their beaches to offload the equipment but before they could do so the Punic fleet was spotted emerging from the harbour at Carthage. Being still loaded, the Roman ships were unable to clear for action and get out to the open sea in time to oppose them. Scipio ordered that they be pulled as close as possible inshore and for the transports to be lashed together to form a wall around them to seaward. The transport's masts and spars were stepped and lashed horizontally across the decks between them, ship to ship, to form a cordon four-deep around the warships; troops were then put on board.[1]

Instead of taking advantage of the situation and attacking immediately, before the Romans had been able to complete their defensive moves, the Punic fleet obligingly anchored off-shore for the night. The next day, the Punic fleet formed up for a formal naval engagement and waited, but the Romans did not emerge and eventually, they attacked what had been made into a floating fortress. The decks of the unladen transports were higher than those of the Punic warships and some of the Roman ship's boats emerged from between the transports to harry the attackers. These however proved ineffective and became a hindrance and were withdrawn.

This initial Punic attack failed to make any headway and their ships thus backed water and withdrew. When they renewed their assault, they concentrated on trying to grapple the transports and pull them out of line to make a breach. This was more successful and despite the Romans best efforts, some sixty of the transports were lost and towed away by the Punic ships. This success was of little consequence as three intact lines of interlocked transports remained and the warships were still secure and undamaged. Punic losses are not known but their attack had failed due to the sheer depth of the Roman preparations.

Cumae 38 BC

In the civil war between Octavian and Sextus Pompeius (38–36 BC), the latter had established himself in Sicily and commanded a fleet of some 130 ships, the larger part of what had been the regular navy's western fleets. Octavian mustered a fleet of nearly ninety ships, being those that had stayed loyal and in home waters and a large number that had changed sides and come over to his cause. The majority of Sextus' fleet had veteran and experienced crews with well worked-up ships; that of Octavian had crews that were mostly either recently recruited and inexperienced, or in the case of the ships that had recently changed sides, probably unreliable (although they proved in practice not to be so).

Learning of Octavian's moves against him, Sextus left forty of his ships at Messina to guard against an attempted crossing from the mainland and moved north with his remaining ninety ships to meet Octavian's fleet, commanded by Calvisius. Sailing south, Calvisius' ships had

View to the south from the acropolis at Cumae towards Cape Miseno. Lake Fusarus is in the foreground and the low-lying island of Procida is in the background at the right. It was here in 38 BC that the fleet of Sextus Pompeius engaged the Octavian fleet under Calvisius, whose ships were forced to back on to this shore and from which he had to fight a defensive, static battle. (*Author's photograph*)

already suffered some damage from a storm and were then met by Sextus' fleet off Cumae and before they could round Cape Miseno, to reach the safety of their harbours in the Bay of Naples. Both fleet formed line of battle abreast, Calvisius with his left flank close inshore. When they met, Sextus' attack started to push Calvisius' ships round and back towards the shore. Calvisius then ordered his ships to back water and pull back to moor stern-on against the shore, close together and where he could draw support from some of Octavian's troops who had been shadowing the fleet.

Sextus renewed his attack with his left wing, setting two ships against one and causing a great deal of damage. This however, had the effect of shortening his line and seeing this, Calvisius, with the ships of his other wing, cut their cables and pulled out to counter-attack the rear of Sextus ships, capturing his flagship and driving the rest off.

Sextus withdrew to Messina and Calvisius with his surviving ships continued down the coast, intending to rendezvous with Octavian at Reggio di Calabria, where the latter had a squadron of warships from the Adriatic. Octavian put to sea and seeing this, Sextus also sallied forth to intercept before the two enemy formations could combine. Octavian was outnumbered and Sextus caught up with him near Cape Pelorus and started to attack from the rear. Rather as Calvisius had done, Octavian pulled into the mainland shore and moored stern-on to the shore to withstand the attack.

Once again, Sextus' ships attacked the end ships of Octavian's line, two to one and started to systematically destroy them. On one occasion, Octavian's own ship was attacked and he was forced to jump overboard and make for the shore. Some of the ships were again able to cut cables and counter-attack, but the weight of their opponent's greater numbers was beginning to turn the battle in Sextus' favour. When he saw Calvisius' fleet approaching and in fear of being caught between them, Sextus broke off the battle and retired. Octavian's ships suffered some more damage from another storm or squall, before they could be cleared from the beach and withdrawn to the safety of their harbour. In all he lost half of his fleet to these actions, despite a spirited and ultimately successful defence, from a tactic that had at least minimised his losses against a clearly superior enemy fleet.[2]

PRINCIPAL ROMAN NAVAL ACTIONS

338 BC Anzio. Romans win their first recorded naval action.

310 BC Pompeii. Roman amphibious landings meet with mixed success.

282 BC Roman squadron defeated by Tarentines in Gulf of Taranto.

260 BC Mylae, Sicily. Romans beat Carthaginian fleet in first major battle of the First Punic War.

258 BC Sulci, Sardinia. Romans beat Punic fleet.

257 BC Tyndaris, Sicily. Romans beat Punic fleet.

256 BC Ecnomus, Sicily. Romans beat Punic fleet.

249 BC Trapani, Sicily. Punic fleet destroys Roman in their only defeat of the war.

249 BC Camarina, Sicily. Storm interrupts battle against Punic fleet.

241 BC Egadi Islands, Sicily. Romans beat Punic fleet to end First Punic War.

217 BC Ebro Delta, Spain. Romans destroy Punic Fleet on the shore.

210 BC Roman squadron defeated by Tarentines off Taranto.

208 BC Roman fleet defeats Punic fleet off African coast.

207 BC Roman fleet again beats Punic fleet off African coast.

206 BC Roman squadron beats Carthaginian squadron in Strait of Gibraltar.

201 BC Romans beat off Punic fleet attack on their fleet while it is against the shore.

191 BC Cape Corycus, Rome and Pergamum beat Seleucid fleet.

190 BC Myonnesus, Rome and Rhodes beat Seleucid fleet.

147 BC Carthage, last sortie of Punic Navy.

56 BC Caesar beats the Veneti, Quiberon Bay, France.

49 BC Marseilles, two civil war battles during siege.

47 BC Alexandria, Egypt. Caesar beats Ptolemy's fleet.

38 BC Two civil war battles off Cumae. Octavian versus Sextus Pompeius

36 BC Milazzo, Agrippa beats Sextus Pompeius.

36 BC Naulochus, Sicily. Agrippa beats Sextus Pompeius.

31 BC Actium, Greece. Octavian and Agrippa beat Antonius and Cleopatra.

AD 268 Goths in the Aegean

AD 323 Dardanelles. Civil war between Constantine and Licinius.

AD 386 Danube, Romans destroy Goth invasion.

AD 456 Western Roman fleet defeats Vandal fleet off Corsica.

AD 467 Roman fleet destroyed near Carthage by Vandal fleet; last action by Western Roman fleet.

PART IX ALLIED NAVIES

27 RHODES

Building tradition

The island of Rhodes had been settled by Dorian Greeks some time prior to the eighth century BC.[1] It was engulfed by the Persian Empire in the mid-sixth century BC and, as a tributary maritime province, contributed to the contingent of 100 Ionian Greek warships that formed part of the Persian fleet accompanying their invasion of Greece, that was defeated by the Greeks at the battle of Salamis in 480 BC.[2] With the Persian defeat at Plataea in 479 BC, the remnant of their fleet retired to Samos,[3] where the Greek fleet caught up with them. The Ionians revolted and the Persians were driven out.[4] Rhodes joined the Athenian-led Delian League, contributing a nominal two penteconters to the Athenian fleet for its ill-fated expedition against Syracuse in 415 BC.[5] Rhodes seceded from the League in 411 BC, setting up a republic and unifying the island by 407 BC.[6] It next sided with the Peloponnesians and contributed ships and men to their fleet. This led to the use by the Spartan fleet of Rhodes' harbours between 398 and 396 BC, amounting to a virtual occupation and led to Rhodes defecting from Spartan hegemony, backed by a Persian fleet. All of this persuaded the island that it needed to be in a league with the other Aegean island and have a navy of its own. By 390 BC, it had a fleet of at least sixteen warships, enough to thwart an eight-ship Spartan attempt against them.

Alliance with Athens brought security but once again, had soured by 357 BC to the extent that an Athenian fleet of sixty ships was twice defeated by a Rhodian and allied fleet, before peace in 355 BC brought formal recognition of Rhodes' independence.[7] Unfortunately it then fell under Carian domination and had to send warships to aid Byzantium against Macedon in 340 BC. With Alexander the Great's march into Asia, Persian hegemony was ended and in early 332 BC, Rhodes sent ten warships to offer submission to Alexander at Tyre. Rhodes received a Macedonian garrison but with the death of Alexander in 332 BC, it was ejected and true independence established. With the waning of Athenian power and influence, Rhodes was excellently placed to take advantage and become the leading trading centre of the eastern Mediterranean and Aegean. Its prominence as a seafaring nation had already been established as early as about 500 BC, when it promulgated the first law for maritime trade, which became widely accepted internationally and some principals of which have continued in use up to the present. The state expanded to include some nearby islands and a large tract of the adjacent mainland. As a trading nation with a large and growing merchant fleet, that trade had to be protected, especially against the ever-present threat of piracy, Rhodes being near to Crete and coasts on the mainland notorious for such activities, the establishment of a navy was a necessity.

Rhodian concern and attention to the maintenance and upkeep of their navy in first class condition was paramount, the success of the state depending upon it. The crews were recruited from the citizen body (the army depended on mercenaries to a large extent). The navy was of between thirty and forty warships, with a basic tactical unit of three ships, commanded by an *archon*. In wartime the fleet was commanded by a *navarch*. The fleet had a dedicated military harbour with shipbuilding yards, dockyard and facilities adjacent to Rhodes city.[8] What they may have lacked in numbers, the Rhodian navy more than made up for in quality, as a review of its actions testifies.

The Rhodians eschewed the large Hellenistic ships and concentrated on smaller types, but which were modern, well maintained and manned by the finest, well trained crews; their skill at swift manoeuvre and ramming became famous. Lacking the numbers to be able to engage in the grappling and boarding tactics of other powers, the Rhodians relied on their superior seamanship and ship handling. They operated quadriremes as their largest type (although quinqueremes would be added later) augmented by triremes and a type of which they were particularly fond, the trihemiolia. The exact nature of this type is not clear but, meaning 'three and a half' perhaps it was a form of trireme with half of the thranite oars double-banked; perhaps a bireme with an added half reme at the thalamite level (would this however be a 'two and a half'?). It has been interpreted[9] as a trireme with the two topmost remes rowing through an oarbox and a half reme at the lowest, thalamite lavel, for a total of 120 oars, each single-manned. However, this would appear to produce an under-powered trireme with no obvious advantage. Whatever the form, the type was intended as a ramming vehicle, beamier and better protected than the trireme.

Enriched by its trade and industry, which boomed after Alexander's demise and aided by their navy's continuous and strenuous action in suppressing piracy, Rhodes built up the largest merchant fleet in the eastern Mediterranean. In 306 BC, the island concluded a treaty of friendship with Rome.[10] This success made Rhodes a tempting target and Antigonus I (306–301 BC) who ruled much of Alexander's Asian conquests and who sought maritime supremacy, sent his son, Demetrius Poliorcetes ('the besieger', 306–283 BC) to lay siege to the city. Faced by a fleet of 200 warships, joined by many pirate ships, the Rhodian fleet withdrew into its harbours but some ships slipped out to interdict Demetrius' supply ships and take prizes.

Demetrius mounted an attack from the sea with ships roped together to provide armoured platforms for assault troops and other with four storey towers and artillery. The first assault at night secured part of the main harbour mole. The next day ships mounting catapults got into the harbour and did great damage; they were withdrawn at night, when the Rhodians came out in boats and set fire to most of them. Further attacks from seaward were eventually

beaten off. In one such, fire ships were sent into the harbour; three Rhodian ships sallied out and broke through Demetrius' boom and sank two of his artillery vessels. Unfortunately they ventured too far and were counter-attacked and rammed, one being captured by the enemy. In bad weather, an attack overran the enemy troops on the mole. Reinforcements were received from Ptolemy of Egypt and the Rhodians sent three squadrons of three ships to attack Demetrius' supply lines, which they did with great success, bringing captured supplies into the city. Demetrius' fleet sailed as a diversion for massive land attacks, only to be beaten off with great difficulty and the help of more reinforcements from Ptolemy. Finally, in 304 BC, the siege was called off.[11]

The third century BC, witnessed a rise in Rhodes' prestige and power, with it intervening in Hellenistic affairs on occasion to protect or enhance its trade interests. Rhodes' navy was kept in first class order and pursued a continuous campaign against pirates and freebooters; they were acknowledged as 'the leaders in maritime affairs.'[12] In 228 BC, the island suffered a great earthquake and as part of the rebuilding, funds were applied to build ten quinqueremes.[13]

Ally of Rome

By the late third century BC, the expansionist ambitions of Philip V of Macedon threatened Rhodes and in 208 BC, they joined other Greek states in an embassy to mediate with him;[14] an exercise repeated in the following year. The First Macedonian War ended in 205 BC, with Rhodes' sympathies leaning towards Rome who, having suppressed Illyrian piracy and removed the Carthaginian's power, had provided stable conditions for Rhodes' trade in the West. The Rhodian's concerted action against Crete and its pirates also alienated Philip, as president of the Cretan confederacy.[15] Philip's increasing aggression and build-up of naval power led to open war between them in 202 BC.

The following spring, Philip's fleet seized the Cyclades and Samos. Rhodes sent a fleet of thirty ships to Lade, where they were bested by Philip's greater numbers in a brief battle, losing

The 'Victory of Samothrace': the prow of a Hellenistic warship crowned with a figure of Nike. The prow, although damaged and missing its stempost and the ram, is detailed and shows oarports on both sides. It has been interpreted as a Rhodian dedication and to represent a trihemiolia, one of their favoured warship types of the era. Early second century BC. Louvre. (*Author's photograph*)

two quinqueremes. They did subsequently recover most of the Cyclades. Pergamum, also threatened, now joined Rhodes in the war, their joint fleets meeting and defeating that of Philip at the battle of Chios.[16] A further battle near Miletus resulted in a Pyrrhic victory for Philip and cost him nearly half of his fleet. Rhodes and Pergamum appealed to Rome for help and with the addition of a Roman fleet of thirty-eight ships and those of the allied Athenians, Philip was unable to mount any further challenge at sea. The allied fleets harried enemy coasts, with the Rhodians blockading Philip's ships at Volos (Demetrios) until the end of the war in 197 BC. In 194 BC, Rhodes sent eighteen ships to join those of Pergamum in defeating the bellicose King Nabis of Sparta.

Peace was short-lived as the Seleucid king, Antiochus, started to expand into Asia Minor and then into Greece. The Romans drove him out of Greece but, as Antiochus had a very powerful navy, any further advance would have to be preceded by overcoming it. To the Roman Aegean squadron of twenty-five ships were added another fifty quinqueremes and triremes. Rhodes sent its fleet of twenty-seven ships to join those of Rome and Pergamum in September 191 BC. Outnumbered, Antiochus' fleet withdrew into Ephesus and the Rhodians returned home for the winter.

In April 190 BC, the Rhodian fleet of twenty-seven ships was in harbour at Panormus on Samos, just south of Ephesus. Antiochus' fleet managed to slip out of harbour, landed troops behind the harbour, forcing the Rhodians to try to escape by sea; the enemy fleet was waiting to pick them off as they emerged from the harbour mouth. Seven Rhodian ships managed to escape by suspending braziers of burning material at the end of long spars on either side of their bows, but the other twenty ships were lost or captured.[17] With the Roman and Pergamene

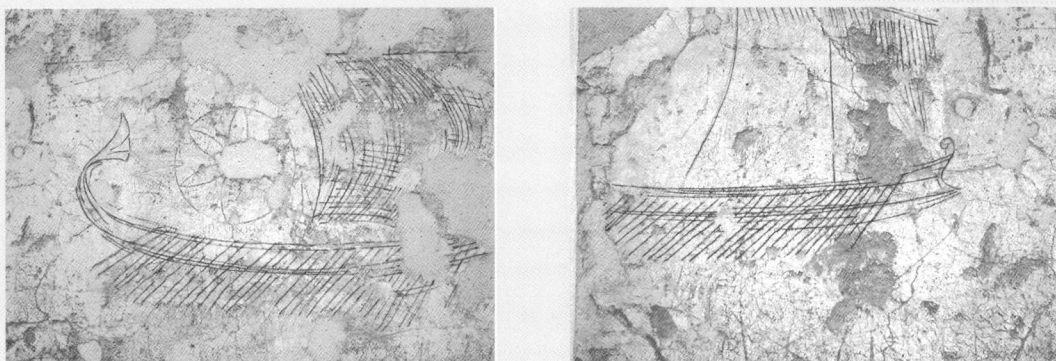

Hellenistic warships under sail and oar. Graffiti scratched into wall plaster at Delos; late third or early second century BC. Although very roughly drawn, they must represent a more or less accurate overall impression of the ships seen by the 'artist'. Oars have been indicated in a random way but the upper ship has an indication of thirty-eight oars and the lower, of twenty-nine, which suggests that they represent multireme ships, but in the absence of any *artemon*, probably not the great polyremes. Delos Archaeological Museum. (*Author's photograph*)

fleets committed to watching Ephesus, it was learned that Hannibal (in Antiochus' service) had been raising another Seleucid fleet in Phoenicia and Syria. Despite their losses, by June 190 BC, Rhodes had managed to assemble a fleet of thirty-two quadriremes and four triremes, with which they sailed to confront Hannibal.

Hannibal had three 'sevens', four 'sixes', thirty quinqueremes and quadriremes and ten triremes, forty-seven ships, mostly larger than those of their opponents. The two fleet met off Side, Hannibal's ships in line abreast, their right to shoreward and Hannibal on the left (seaward) wing.[18] As they could not match Hannibal's line, the Rhodians formed into several short columns of ships in line ahead, their offshore units coming into action first. The columns forced their way in a series of *diekplous* attacks, between and behind the enemy ships, concentrating on the enemy right and centre. The Rhodian right was heavily engaged offshore but the inexperienced Seleucid crews were no match for the speed and manoeuvrability of the Rhodian ships, which overwhelmed their inshore wing. Hannibal was out-matched and ordered a withdrawal, having lost a 'seven' and suffering a dozen more ships damaged and disabled; his fleet was demoralised

and went home, to play no further part in the war. Rhodes kept a squadron of twenty ships to watch the Cilician coast in case they returned.

The Rhodians sent twenty-two ships to join the Roman fleet watching Ephesus, the Pergamene ships having withdrawn. The joint fleet finally met Antiochus' fleet off Myonnesus in October 190 BC. Antiochus' fleet of ninety ships, including two 'sevens' and three 'sixes', formed into line abreast with its right flank to seaward; the Romans fleet of fifty-eight ships formed line with the Rhodians astern of them. The Syrians extended their offshore wing to envelop the smaller number of Roman ships when the Rhodians came out from behind to the seaward flank, using their superior speed and ship-handling, as well as the threat of some fire baskets, to frustrate the enemy's attempt at a *periplous*. They held the end of the enemy line while the Romans broke through the enemy centre and fell upon their rear, then proceeding to methodically grapple and board enemy ships. Before they finally managed to break away and retreat, Antiochus' fleet had lost thirteen ships sunk or burned and another twenty nine captured, nearly half of his fleet. The Romans lost two ships, the Rhodians only one.[19]

With the end of the war in 189 BC, Rhodes's

Third Macedonian War (172–168 BC) they were not needed and returned home. Rhodes sent a squadron of warships to join Rome against Carthage in 147 BC. The peace settlement in Greece of 167 BC affected Rhodes in that the Romans declared the Athenian island of Delos a free port; it quickly grew and the loss of trade sent Rhodes into a gradual decline as a trading nation leading to a reduction in its navy and their ability to police the seas, with a consequent increase in piracy, which became endemic. At the end of the second century BC, the Romans sent several naval expeditions to combat piracy and used Rhodes as a base and were supported by Rhodian ships.

In 89 BC the Pontic king, Mithridates VI sent his fleet into the Aegean, where it sacked Delos, but was checked by the Rhodian fleet. By 86 BC, the Dictator Sulla sent his deputy, Lucullus to obtain a fleet and Rhodes furnished him with three warships, which formed the basis of the allied fleet that he raised to support Sulla in Asia Minor. Rhodian ships were again in action alongside the Roman fleet during the second war against Mithridates (74–73 BC) as well as then providing a base and facilities for Pompeius in his war against the pirates of 67 BC, when that campaign moved east against the pirates' strongholds on the south Turkish coast.

The civil wars of the first century BC, found a Rhodian fleet of sixteen ships in the Adriatic, forming part of Pompeius' fleets opposing Caesar.[20] In April 48 BC the Rhodians were at Durres, Albania, when Antonius's fleet, sailing with supplies and reinforcements for Caesar was caught by strong winds and blown past. The Rhodians emerged to intercept, but Antonius' ships managed to gain harbour, leaving the Rhodians exposed to a sudden squall which forced their ships on to a rocky shore; the ships were lost and the survivors of their crews captured by Antonius.[21]

After the defeat of Pompeius, Rhodes quickly realigned itself with the victorious (and forgiving) Caesar and furnished ten warships to add to his own few ships when he went to Egypt, one being detached to get reinforcements.[22] Trapped with Caesar in Alexandria in 47 BC, it was the Rhodian ships that sallied to beat off the Egyptian fleet trying to intercept Caesar's reinforcements, rescuing one of their own ships when it was assailed by four of the enemy.

With the reinforcements, Caesar's fleet was now of nine Rhodian and twenty-five Roman ships, all placed under the command of the Rhodian *navarchos* Euphranor. He led the fleet out of the eastern harbour of Alexandria in a bold attack on the Egyptian fleet of twenty-nine ships, plus some smaller, which was in the western harbour. The entrance was narrow, forcing Euphranor's ships to enter singly, nevertheless they gradually forced the Egyptians back and disabled and captured several of their ships, the remainder fleeing to the safety of a nearby canal. The fleet was again in action beating off an Egyptian attack on a supply convoy. Another Roman fleet was now advancing from the east and Caesar sailed out on the Roman/Rhodian fleet to join them for the actions that ended the war.[23]

In the civil war following Caesar's death in 44 BC, one of the anti-Caesar conspirators, Cassius became commander of the Roman eastern fleet. Rhodes, unsure of the situation, refused to acknowledge his authority and Cassius attacked. The Rhodian fleet of thirty-three ships sallied but were beaten by Cassius' eighty ships,[24] losing two sunk and three captured. A later sally cost the Rhodians two more ships. Cassius prepared to besiege Rhodes city, but it was entered and fell without the need to do so.[25]

Rhodes continued, notionally independent, but as a part of the Roman world until Claudius annexed it to the province of Asia in AD 44. It continued to furnish a few nominal ships to sail as an adjunct to those of the Imperial fleets and two were recorded a late as the reign of Titus (AD 79–81).[26]

28 PERGAMUM AND SYRACUSE

Pergamum

Pergamum (near modern Bergama in western Turkey) and its surrounding territory of Mysia had formed part of the empire of Seleucus, who had made himself master of Alexander the Great's empire in Asia. Frontier wars between Seleucus and the other major successor dynasties, all of whom were fomenting trouble in each other's lands, made it impossible for him to hold the vast disparate territories together. Parts of the empire were able to secede and under its governor, Philetairos (governed 283–263 BC) Pergamum sought greater autonomy; under his nephew and successor, Eumenes (reigned 263–241 BC) an independent kingdom was formally established. The Seleucid king Antiochus I sent an army against him, which Eumenes defeated.[1] Eumenes' family, the Attalids continued to rule the kingdom until 133 BC.

To guard their independence, the Pergamenes built and maintained and efficient navy of about forty modern ships. The composition of the fleet is not known in detail, but at Chios in 201 BC, their fleet included at least two quinqueremes and two quadriremes and was considered to be the heavier division, the allied Rhodian division having lighter ships; to stand in the line of battle, as they did, the fleet must have had a preponderance of larger, heavier ships.[2]

The kingdom first came into the Roman orbit in 211 BC, when they joined the Roman and Aetolian coalition of states formed to contain Philip V of Macedon in the First Macedonian War. King Attalus of Pergamum (reigned 241–197 BC[3]) made an unsuccessful incursion into Thessaly but in 210 BC was appointed *strategos* (general) of the Aetolian League[4] and his fleet, with their Rhodian allies, maintained dominance over the Aegean. Having destroyed Philip's fleet in Illyria, a Roman fleet of twenty-five quinqueremes moved into the Aegean in 208 BC to join Attalus' thirty-five ships at

Lemnos. Philip's Aegean fleet was completely outmatched and could not oppose attacks on his coastal towns by the Pergamene and Roman ships. Faced by problems at home from barbarian incursions, Attalus had to withdraw his fleet in 207 BC.[5]

Philip built a new fleet and in 201 BC annexed the Cyclades and crossed into Asia Minor and attacked Pergamene territory. Pergamum and Rhodes declared war and sent an embassy to Rome to seek help. Philip meanwhile, seized Samos and Chios and landed at Elaea, the port of Pergamum and sent a force to besiege the city itself. Attalus was with his fleet, totalling, with his Rhodian allies, some sixty-five ships. Philip had fifty-three heavy and about 150 light, open ships and as he could not press the siege, he withdrew and sailed to join the rest of his fleet, laying at Samos.[6] Before he could however, the Pergamene and Rhodian fleets caught Philip off Chios and in the ensuing sea battle, Philip's fleet was beaten, losing a 'ten', a 'nine' a 'seven', a sexteres and ten other large ships, as well as twenty-eight light ships. Attalus lost his flagship, two quinqueremes and one other ship, the Rhodians lost three.[7] Despite suffering a reverse near Miletus, the allied fleet had inflicted appalling losses on Philip's fleet, which withdrew homeward. A Roman fleet of thirty-eight ships joined its allies to dominate the Aegean, Philip's fleet making no further challenge and being surrendered at the end of the war in 197 BC.

Antiochus III (reigned 223–187 BC), although carefully bypassing Pergamene lands, seized Ephesus in 197 BC; the city lay in a small enclave separating Pergamene from Rhodian territories. King Nabis of Sparta moved in 194 BC to ally himself with Antiochus, but was defeated on land and an allied fleet of Roman, Pergamene and Rhodian ships attacked his naval base at Gythion, landing troops there. The new king of Pergamum, Attalus' son Eumenes II (reigned 197–169 BC) sent his fleet of twenty-five ships, mostly quinqueremes, to join the Roman fleet of seventy-five ships at Aegina. The allied fleet met with Antiochus' fleet of seventy ships off Cape Corycus in September 191 BC, Eumenes' fleet being in the rear of the Roman ships. The Roman downed sails and cleared for action, deploying into line abreast of about thirty ships (i.e. a double line) and slowed to allow Eumenes' to bring his ships into line to seaward. This proved critical as, when the Romans advanced to general engagement with Antiochus' front, Eumenes was able to round the enemy's offshore flank in a *periplous* manoeuvre. Antiochus fleet had to break off the engagement; the Pergamenes had lost no ships. As it was the end of the sailing season, Eumenes' fleet returned home, the Roman fleet also wintering in Pergamene territory.

In the campaign of 190 BC, seven Pergamene ships went with a Roman squadron of thirty ships to clear the Dardanelles of some hostile shipping. Antiochus' forces attacked Pergamum by land, forcing the withdrawal of their fleet to home waters. Relieved by the arrival of allied Greek troops and with the fleet of Antiochus shadowed by the Roman and Rhodian fleets, Eumenes' fleet went to the Dardanelles to protect the crossing by a Roman army. They were still there when, in October 190, the Roman and Rhodian fleets broke Antiochus' naval power at the battle of Myonnesus.

In the Third Macedonian War the Macedonian king, Perseus (reigned 179–168 BC), had only a small fleet. Once more a Roman fleet of thirty-eight ships was joined by Pergamene and other allied ships but, in the absence of opposition, the allies were not needed and returned home. In 169 BC, however, twenty Pergamene ships joined the Roman fleet and they attacked Perseus' naval base at Cassadeia; the allies then returned south, raiding the enemy coast as they went, it being late in the season. Perseus' ships were not entirely inactive and a fleet of forty light ships freed a convoy of Macedonian grain ships from their Pergamene captors; this of course implies that the Pergamene fleet had been active in intercepting Macedonian trade. In 168 BC, renewed allied operations cleared the sea until the war was concluded in June by the battle of Pydna.

For the remainder of its existence, the Pergamene navy was at relative peace. As a result of the defeat of Antiochus, the kingdom had gained large amounts of territory and coastline, including part of the Hellespont. This far longer coast required a constant watch to counter the ever-present pirates. Internal unrest and dynastic problems led the King Attalus III (reigned 139–133 BC) to bequeath his kingdom to Rome. Disputed by his half-brother, Aristonicus (Eumenes II) the kingdom was finally acquired by Rome in 130 BC and made into the province of Asia;[8] of its fleet, no more is heard but some of its laid-up ships undoubtedly found their way into pirate hands.

Syracuse

Founded in approximately 740 BC by Greeks from Corinth, Syracuse enjoys the larger of the only two natural harbours on the east coast of Sicily. Served by a fertile hinterland, the city prospered and grew. In 485 BC the city was captured by Gelon, successor to Hippocrates of Gela who had subjugated most of eastern Sicily.[9] Gelon made the city his capital and commenced to build up his forces, particularly a fleet in anticipation of war against the Carthaginians of western Sicily. The extent of his efforts can be gauged by his offer in 480 BC, to send 200 triremes to aid Greece against the Persian invasion.[10] Whether he could actually have had such a number is open to question however. In 474 BC, the tyrant Hieron sent a fleet to join that of Cumae in Campania; the allies defeated the Etruscan fleet, effectively ending their sea power. A Syracusan fleet again raided the Etruscan coast in 453 BC. The Syracusan fleet was withdrawn to the safety of their inner harbour following a defeat of their eighty ships by sixty-five Athenian ships. During the final stage of the Athenian siege of 414 to 413 BC, however, it sallied forth to inflict a decisive defeat on the Athenian fleet in the Great Harbour or Bay of Syracuse.[11]

Syracuse. View south from the Achradina district with its cliffs. To the right is the entrance to the smaller harbour, the former base of the Syracusan navy. Beyond is the seaward coast of Ortygia island and in the distance, beyond the end of the island, the coast on the other side of the entrance to the Great Harbour. The approach from the east, is strewn with rocks. (*Author's photograph*)

Either under threat from or in conflict against Carthaginians for hegemony over the island, the tyrant Dionysius gathered leading military engineers and innovators to boost his forces. In 398 BC they had launched the 'five' or quinquereme, as well as developing artillery and building quadriremes.[12] With the building of these bigger ships, Syracuse could deploy a navy of nearly 200 ships, including transports and supply ships, for its siege of Motya in 398 BC.[13] Dionysius II (367–344 BC) introduced the sexteres into service.[14] Between 270 and 265 BC, the tyrant Hiero II (306–215 BC) built a giant freighter at Syracuse, believed to have been over 200 feet (62 m.) in length and capable of carrying some 2000 tons of cargo.[15]

At the start of the First Punic War (264 BC) the Carthaginians managed to gain a rather unnatural alliance with Syracuse, denying its harbour to the Romans. The Syracusan fleet made no attempt however to interfere with the Roman crossings to the island and in 262 BC, ended the alliance and Hiero thereupon became a firm ally of Rome. Roman ships could and did use the city as a base, as they did for example, in 256 BC in preparation for their attack on Africa.

The treaty of friendship between Syracuse and Rome was renewed in 248 BC. There is no record of Syracusan warships either undertaking belligerent activities on their own or operating with Roman fleet during the war, but the use of their harbour and facilities was invaluable to the Romans. At the beginning of the Second Punic War in 218 BC, Hiero sent twelve of his warships, which captured three Punic quinqueremes that had become detached from their fleet and swept into the Straits of Messina. They were handed over to the Romans at Messina and valuable intelligence was obtained from their crews.[16] Two years later, Hiero sent a fleet to Ostia with archers and slingers for the Roman army and a large cargo of grain.[17] A Punic squadron raided the Syracusan coast in that same year.

The size of the Syracusan fleet at this time is a matter for some conjecture. In 440 BC, it is said that they built 100 triremes, however, such a fleet would require approximately 20,000 men to crew them and yet the year before, in a major battle against the other leading power in Sicily, Agrigento (Akragas) Syracuse had fielded an army estimated at 20,000 troops, for a battle on which the survival of the city depended.[18] The number of triremes is thus likely to have been exaggerated and to be to replace old and worn-out ships. During the Athenian siege, it averaged, with allies, between seventy and eighty triremes. By the time of the Punic Wars, the Syracusan navy, once so innovative and technically advanced, was clearly much reduced and past its best and certainly incapable of mounting any serious opposition to the Roman fleet.[19]

Syracuse, the Porto Piccolo, the home base of the Syracusan navy. In the centre of the photograph can be seen the modern road bridge across what was the channel between Ortygia (on the left) and the mainland, linking this with the Great Harbour beyond. (*Author's photograph*)

Hiero died suddenly in 215 BC. He was succeeded by his young son, who had anti-Roman sympathies and took a Roman naval patrol along the south coast of Sicily as an affront. A pro-Carthaginian faction took advantage of the unsettled situation, killed Hiero's son and seized power, forming an alliance with Carthage and declared war against Rome. The loss of such an important harbour, together with the threat of a ready foothold for Punic forces, or even Hannibal, on Sicily could not be countenanced and the Romans immediately laid siege to the city (214 BC). A Punic fleet landed troops on the south coast and fifty-five ships ran into Syracuse harbour but as the garrison could not feed them for long, they left after a short while.[20] The Syracusan fleet was confined to the inner harbour and no attempt was made by it, even with the Punic ships, to challenge the Roman fleet that was closing in on them and numbering about 130 ships. The Punic landing force was defeated and withdrawn.

The Romans decided upon a naval assault against the city, using sixty quinqueremes, most loaded with archers, slingers, javelinmen and artillery to provide cover. Eight of the ships were lashed in pairs and fitted with *sambucae* or lyres, a protected scaling ladder long enough to reach the tops of the walls; they were hoisted between kingposts on the ships, atop of which were platforms for archers. On approaching the walls, the ladders were raised by pulleys and the assault troops would swarm up them, covered by screens and on to the walls. Other ships had towers, battering rams and hooks to tear at the battlements. The assault was to be against the city walls fronting the Great Harbour but the engineer Archimedes had prepared the defences, with loopholes for artillery at low level, covering the usual blind spot (as ancient artillery could not be depressed). Masses of artillery of all sizes had been made ready, as well as counter-weighted swing beams to grapple and lift the ships close to the walls, then drop them, causing swamping and severe damage. There were also 'dolphins', swing beams with a very heavy weight, swung out over the ship and dropped on to the deck, then raised and the action repeated, in the manner of a wrecking ball.[21]

The Roman attack was repulsed, but the siege was tightened until finally, in spring 212 BC, part of the defences were captured. A sally by the Syracusan fleet, its last foray, was easily beaten back and took no further part in matters until the city finally fell in 211 BC. Syracuse lost its independence and became a Roman city, whatever remained of its fleet being absorbed into that of the Romans.

Marseilles

Founded in about 600 BC by Greeks from Phocaea in Asia Minor, who used penteconters for their voyaging, according to Herodotus,[1] the city state of Massalia grew into a prosperous trading centre. After initial battles to secure a hinterland, they established a fortified settlement at the mouth of the River Rhone and then continued to found others along the southern French coast. This brought them into conflict with Carthaginian interests as they pursued a policy of setting up other posts along the Spanish coast. An attempt to expand into Corsica was thwarted in 535 BC when their fleet of sixty warships was narrowly beaten by a combined Carthaginian and Etruscan fleet.[2] Their position between the Carthaginians and Etruscans meant that a war fleet had to be maintained by the city and they survived by winning two naval battles against the Carthaginians. A treaty with Carthage in 490 BC fixed the boundaries of their respective spheres of interest at Denia, a little south of Valencia.[3] Carthaginian expansion continued, edging the Massaliots out of Spain and dominating the Western Mediterranean. In their isolated position and under such threat, the city turned the growing power of Rome as an ally.

They made common cause when the Ligurian tribes advanced into Etruscan territory and engaged in piracy, leading to Roman military campaigns against them between 238 and 230 BC. The nature of the terrain, with mountains coming down to the sea and no lateral roads, dictated a mainly naval campaign, essentially a series of amphibious landings and it is reasonable to assume that Massaliot ships assisted the Romans and provided them with a secure base on the other, non-Roman side of the area of operations. The Roman sea coast was extended to La Spezia (Luna) as a result of these campaigns.

Two Massaliot warships were scouting for the Roman fleet in Spain in 217 BC and it was their efforts that enabled the Romans to engage and defeat the Punic fleet of Spain in the battle of the Ebro.[4] By 161 BC the Romans had advanced their border to meet that of Massalia and they extended their road from Pisa (Pisae), through Genoa (Genua) to the city itself. The route was not yet secure and in 154 BC, Ligurians raided Massaliot ports at Antibes (Antipolis) and Nice (Nikea). The Romans marched and defeated the Ligurians, securing the road to Massalia which effectively became a client state. This status did enable the Massaliots to seek Roman help in 125 BC against tribes of Gauls and Ligurians. The subsequent Roman campaign led to the creation of the province of Gallia Narbonensis, i.e. the South of France and leaving Massalia as an independent enclave within Roman territory.

The final act for independent Massalia came when she sided with Pompeius in his civil war against Caesar. In April 49 BC the city was isolated by Caesar's forces and under siege from landward and blockaded by a fleet of twelve warships, including a sexteres, under Decimus Brutus. In preparation, the Massaliots had refitted such warships as they could obtain, namely a quinquereme, two triremes and three bireme liburnians; these were augmented by eleven large merchant ships, adapted by making provision for oars and fitting of catapults and protective decks, other small boats were also manned by archers and marines.[5] In their first attempt to break the blockade, the Massaliot fleet sailed out in line abreast and initially inflicted casualties on Brutus' men with artillery and archery shooting. Brutus' ships and men were of superior quality however and the Massaliots soon found themselves outclassed and pushed back towards the shore. By grappling and boarding, they took six of the Massaliot ships and destroyed another three, the rest being driven back into harbour; Brutus had lost no ships.[6]

Cut off from landward, the city was doomed if they could not break the blockade and so once more, they set about forming a fleet to challenge Brutus. They managed to fit out some more ships and manned open fishing boats with archers.[7] In the interim, Pompeius had sent a squadron of sixteen warships under Lucius Nasidius to relieve the city; he had surprised and captured a laid-up warship at Messina and sent a scout ship ahead to alert the Massiliots of his approach. The latter managed to evade Brutus and effect a rendezvous with Nasidius' fleet east of the city. Brutus had refitted the six captured ships and he sailed to meet the combined enemy fleet. The latter formed for battle with the Massaliot ships on the right, i.e. to landward and Nasidius' ships on the left, to seaward. When battle was joined, the Massaliots fought with desperation, whereas Nasidius' men, far from home and without the same impetus, only put up a token fight before pulling out of line and sailing for Spain. Despite fighting ferociously, the Massaliots were overwhelmed and lost four ships destroyed, one capsized and four more captured; Brutus lost one ship.[8] By July, Caesar returned and the city, beaten at sea and reduced by the siege, surrendered to him, ending its independence.

Judaea

Rome's first diplomatic contact with Judaea came in 161 BC, when an embassy was received from their leader, Judas Maccabeus and a treaty signed, promising mutual help if either were attacked.[9] The next official contact between Rome and Judaea came with the end of the pirate wars in 66 BC and Pompeius' final defeat of Mithridates in 66 BC, which had left Syria and Judaea in a state of unrest. Pompeius settled affairs in the Levant by annexing both lands as provinces, in 40 BC however, Judaea was converted into an independent client kingdom, ruled by Herod (Herod the Great, reigned 37–4 BC) a status that would continue until the death of Herod Agrippa in AD 44, when it reverted to the status of a province.[10]

There is no evidence the Herod had a navy, but he did in 16/15 BC himself voyage to join Agrippa and his fleet. The king sailed via Rhodes to join him and caught up with Agrippa at Sinop (Sinope) on the north Turkish coast. Although helping Agrippa, there is no mention of him having ships or forces, but, having been of help, it must be assumed that he did at least have at least some, carried by his own ships.[11] His successor, Archilaus (reigned 4 BC to AD 6) issued coins showing a warship and he may well have had few ships although the proximity of imperial fleets, at Antioch and Alexandria made their usefulness questionable.[12]

30   CARTHAGE, TARANTO, THE PIRATES

Carthage

Carthage was founded in 814 BC by Phoenician (Poeni in Latin, hence Punic, meaning Carthaginian) traders from Tyre, on a peninsula projecting into the Gulf of Tunis. As a trading colony, the young city flourished but with the conquest of Phoenicia by the Assyrians in the late eighth century BC, the link with its homeland was broken and Carthage, together with the other Phoenician colonies in the west became independent entities. Carthage founded its first colony on Ibiza in 654 BC and after initial reverses, in alliance with the Etruscans, blocked Greek expansion in the West. This, together with waning Etruscan influence, allowed Carthaginian trading ships free reign in the central and western Mediterranean, including visiting the young Rome. In 508 BC they entered into a treaty of friendship with Rome, a provision of which was that the ships of Rome and her allies were forbidden to enter African waters west of Cape Bon, except in emergency (and even then, had to leave in five days).[1]

In 480 BC Carthage lost a part of its invasion fleet bound for Sicily, to a violent storm, an occurrence that would later afflict the Romans. In the same year they were badly defeated by the Sicilian Greeks at the battle of Himera, after which, Carthage turned its attention to its own African hinterland. Most of the fifth century BC was spent by Carthage in extending its territory there and in subjugating the native peoples there and also in acquiring dominance over the other Phoenician settlements in North Africa and Sardinia.[2]

The fourth century BC was marked by almost constant warfare between Carthage and the Sicilian Greeks over mastery of that island. In 409 BC, the Carthaginians developed and introduced the quadrireme into their fleets. In 348 BC, a second treaty was signed between Rome and Carthage, which was even more restrictive of the Romans, excluding them from Spain and even Sardinia, but allowing trading privelages.[3] A third and final treaty was made in 306 BC on more equable terms, the Romans agreeing not to intervene in Sicily and Carthage to keep out of Italy; they agreed to divide Corsica.[4] Carthage by now had an empire comprising most of modern Tunisia and part of the coasts of Libya, Algeria and Morocco; it held a part of southern Spain, the Balearic Isles, Malta and nearly half of Corsica and Sardinia. Western Sicily was retained in the face of Greek hostility.

Carthage was above all a seaborne trading empire, secured by a navy that was the largest in and dominated the central and western Mediterranean. During the Pyrrhic War, in 278 BC, Carthage sent a fleet of 120 warships to Ostia 'to help'. Pyrrhus was by then already in Sicily and the point of the exercise seems to have been to remind the Romans of Punic naval might. They did however transport 500 Roman troops to augment the garrison of Reggio. In 272 BC a Punic fleet appeared off Taranto during the Roman siege, but sailed off, adding to the growing tension between the two powers.

The mercantile harbour of Carthage had been excavated to provide a huge rectangular basin, at one end of which was a passage to the open sea; at the other end of the basin, a canal led to another excavated basin, circular in shape with an island in the centre, this was the military harbour. All were protected by walls. The military harbour and its island had slipways and ship sheds far hauling the warships out of the water, together with building and repair yards, stores, workshops, in fact everything needed for the fleet. The harbour could accommodate some 200 warships.[5] With the ships stationed across their domains, the Punic navy had a total of about 250 ships and included a 'seven' as flagship, some *sexteres*

Carthage, the military harbour, showing the outlet to the sea. The shoreline has changed since ancient times and no trace remains of the installations and walls that formerly stood here. The channel probably started as a small, covered conduit to allow a flow of seawater to help keep the otherwise almost enclosed harbour from silting. It was cut out during the final phase of the siege in 146 BC, to permit a sally by the last Punic fleet. (*Author's photograph*)

and numerous quadriremes and triremes; the backbone of their fleet was the quinquereme. They maintained about a hundred ships on active service and could commission extra ships from the remainder as needed. With their long experience and practice, the Carthaginians preferred ramming tactics.

An important difference between Rome and Carthage was the availability of manpower. For Rome, its citizen levy and the levies of its allies made every able-bodied man liable for military service; they were free men, citizens or from allies who had rights and status. Carthaginian citizens were largely exempt from military service and although providing officers and a core of citizen soldiery, were not subject to the universal liability for military service that made the whole of the Roman and allied male population potentially available. For the rest of their forces, the Carthaginians had to rely on conscripting drafts of men from their subject peoples and hiring mercenaries.[6] This system set a financial and numerical limit on manpower, a difference that would also become manifest in that whereas in the Second Punic War, Hannibal would prove unable to foment any major desertion of the Italians against Rome, she in turn was able to provoke open

rebellion among Carthage's subject peoples, against which they had to direct a substantial proportion of their military effort.

Despite this, Carthage was able to field substantial forces, most of which did remain loyal and at the start of the Punic Wars Polybius said of their navy 'the Carthaginians naturally are superior at sea both in efficiency and equipment, because seamanship has long been their national craft and they busy themselves with the sea more than any other people.'[7] At the beginning of the wars, the Punic fleet undertook raiding cruises along the Sicilian and Italian coasts, with limited effect. Although they had suffered naval defeats before, the battle of Mylae in 260 BC and the signal defeat that they suffered had an effect beyond the mere losses of men and ships, all of which could be replaced. Their confidence was broken by the scale of the defeat and they surrendered the initiative to the Romans, standing idly by while the latter occupied Corsica and most of Sardinia. Successive defeats at the hands of the Romans showed that although still numerically powerful, the Punic Navy could not match the fighting ability of its enemy.

That the Punic navy was able to effect competent strategy is demonstrated by their

Carthage, the view from the canal from the civilian harbour into the basin of the military harbour. Larger in ancient times, this circular harbour was lined with slipways, shipyards and shipsheds for the building and upkeep of warships. On the island in the centre of the picture, were more shipsheds and slips and the headquarters of the Punic navy. The harbour could accommodate some 200 warships plus all of their gear.
(*Author's photograph*)

dispositions prior to the battle of Ecnomus in 256 BC, where they made excellent use of the local topography, keeping their fleet hidden, ready-formed in its battle squadrons but behind a headland from which their lookouts could signal the Roman's approach. Again, at Trapani in 249 BC, the quick-witted response of the Punic admiral took full advantage of the unbelievable incompetence of the Roman Consul. Too many times however, Punic fleets were caught by the Romans and simply out-fought, as at Cape Bon in 256 BC. By the end of the first war, the Romans had lost some 200 ships in combat but had inflicted losses of 450 ships on the Carthaginians; their navy was reduced to five quinqueremes, four quadriremes, five triremes and some smaller ships by the war's end.

For the Second Punic War, the Romans continued to retain the initiative at sea. Although Carthage had rebuilt its navy to a size that could have challenged the Romans, it failed to do so, with the initial result that Hannibal had to march to Italy, rather than sail. Whereas the Romans maintained their fleet at the highest level, despite the terrible losses suffered in battles, Carthage failed to try to take the initiative at sea, allowing the Romans to dictate the course of the war. Even more than in the first war, this proved to

be vital as the theatres of operations multiplied. In Spain the Carthaginian navy was beaten early on (battle of the Ebro 217 BC) and never again attempted to challenge the Romans there despite the value of that country's resources to them. The allowed the Romans to effectively seal off and contain the Macedonian War, where effective Punic fleet action could have enabled Philip to join Hannibal, as had been intended, a move which could have proved fatal for Rome. Above all, the Punic navy missed perhaps its greatest opportunity when the Romans were tied down at their siege of Syracuse between 214 and 211 BC.

At the end of the second war, Carthage was allowed to retain only ten triremes, the remainder of its ships being towed out to sea and burnt. For the Third Punic War (151–146 BC) with Carthage itself besieged by the Romans, the city managed to assemble a scratch fleet of fifty warships. By cutting a channel from their military harbour to the sea, they were able to emerge in an attempt to break the siege and take the Romans by surprise. Incredibly, instead of taking advantage of this and attacking immediately, the Punic fleet spent three days cruising in the Gulf of Tunis to work up, losing their only advantage. Obviously the veteran

Roman crews prepared and brought them to battle, routing them back against the city's sea walls. The Roman marines in driving their enemy on to the shore, followed and after some heavy fighting, took and secured a large section of the seaward side of the harbour, sealing the city and its fate.[8]

Mediterranean pirates

Although not a state or a naval power in the conventional sense, piracy in the Mediterranean grew to a level that, in 67 BC, caused the Roman Senate to pass a law, the *lex Gabinia*, the sole aim of which was to concentrate the empire's military resources upon driving it from the seas. Piracy had always been endemic (Jason and the Argonauts after all, had set out to steal the Golden Fleece) and in an age before defined territorial waters and organised navies, it had been a normal feature of seagoing. A ship setting out on a cruise could as easily do some buccaneering if opportunity arose, it was 'trade by other means' and could even be condoned by a local ruler, doubtless for a suitable percentage.[9] With the coming of the Persian Empire and the opposing Greek states, large national navies came into being and absorbed all available seamen. Times of peace however corresponded with an increase in piracy. The Athenian Empire, wholly dependent as it was on seaborne commerce, policed the trade routes and kept piracy in check. Later Rhodes, another nation dependent on seaborne trade, built and maintained a substantial navy of excellent quality to suppress piracy and also used armed merchant ships. Rhodes was particularly active due to its proximity to Crete, one of the most notorious and active homes to widespread piracy; also astride its trade routes to the Levant, was the Cilician coast (south-west Turkey) another favourite haunt.[10]

With the succession of wars against Macedon, Seleucia and Pontus, Roman and allied fleets operating in the Aegean and eastern Mediterranean kept piracy in check if not actually fighting privateers serving their enemies. Thereafter piracy grew apace. Pirates had preferred small, swift, easily hidden vessels, fast enough to catch their prey and capable of carrying enough men to overpower the crew of a merchantman. Monoreme conter vessels of from ten to twenty-five oars per side, with a sailing rig that could be lowered; a ram was unnecessary, their objective being capture, rather than damage. A few ships did have rams and local boat types were used, such as the Illyrian *lembi*. Cilician pirates later used 'twos' and 'threes' although it is not clear whether these were biremes and triremes or that rowers were double or even triple-banked at each oar. After the wars however, there were a number of warships of the defeated powers lying in harbours or shipyards which, together with the men who knew how to operate them, added massively to the pirate's capability. Some of the vessels used by the pirates in fact became the prototypes for craft developed and adopted by regular navies such as the *myoparo*, the *hemiolia* used by the Rhodians and most notably the *lembus* and *pristis* of the Romans, which became the *liburna*.

With hundreds of ships from small, open boats to major warships, perhaps even a few Hellenistic polyremes, by the seventies BC, it was said that there were a thousand pirate ships. It was perhaps little wonder that the successive Roman expeditions to suppress them met with very limited success and even occasional defeat. Expansion of Roman territory in Illyria, the occupation of the Balearic Islands in 123 BC and especially the taking of Crete in 68 BC, reduced the pirate's bases of operation. They were resilient however and re-emerged elsewhere until the full force of the Empire had to be concentrated against them by Pompeius in his war against the pirates of 67 BC, which left the Mediterranean free from piracy for centuries to come.

Taranto

Taranto, Taras in Greek and Tarentum to the Romans, was founded by Greek settlers in the late eighth century BC. By the mid-fourth century BC, the city was seeking to increase its domains in Apulia, enlisting the help of King Alexander of Epirus.[11] Although the Roman coastline and territory would not approach their area of interest for some years to come, the Tarentines were sufficiently concerned at the possibility of Rome being tempted to interfere or intrude by sea that they entered into a treaty with her in 338 BC. By this treaty, the Romans agreed that their warships would not sail east of the Capo Colonna, the Lacinian promontory near Crotone in the western Gulf of Taranto.[12]

After the Third Samnite War (ending in 290 BC) the Romans has acquired an Adriatic seaboard as far south as Bari. This proximity and their alliance with the Apulians provoked tension and relations reached breaking point in 282 BC; the Greek port of Thurii in the western Gulf of Taranto, appealed to the Romans for help. Well placed to act as a base between the Tyrrhenian and Adriatic seas, the Romans responded by sending a squadron of ten warships into the Gulf. Whether being deliberately provocative or not and arguably in breach of the earlier treaty, the Roman ships were cruising near to Taranto, which responded by sending its fleet of between fifteen and twenty ships into the attack. The Tarentine attack sank four Roman ships and the rest scattered and ran; the Roman's first naval defeat.

The Tarentines proceeded to sack Thurii and refused Roman peace envoys, leading to war. As the Romans marched, the Tarentines appealed once more to Epirus for help, which arrived in the person of King, Pyrrhus and his army. The ensuing war (280–275 BC) saw no naval battles and with the departure of Pyrrhus, the Romans closed in and Taranto surrendered in 272 BC, its ships being rated as allied and able to be added to Roman naval strength as required, although with what dependability is questionable.

Taranto was to remain an uneasy part of the Roman world until the Second Punic War. Hannibal and his army had failed in a surprise move on the city in 214 BC, when the Roman fleet arrived to reinforce and support the garrison. In the following year, the Tarentines were angered by the Roman execution of some hostages and opened their gates to Hannibal, who was eager to gain a safe harbour as a base. The Roman garrison however, withdrew to the citadel which dominated the narrow harbour entrance (now closed by a road bridge) and held out, supplied by sea. Hannibal could only build a ditch and palisade around the citadel to pen the Romans in and, denied his harbour, had to drag ships from the inner harbour, across the isthmus (ironically where the canal now is) to get to sea.[13]

In 210 BC the garrison was becoming short of supplies and a provisioning fleet with an escort of two triremes, three quinqueremes and twelve smaller ships from Reggio di Calabria, Velia and Paestum. Learning of their approach, the Tarentines managed to drag about twenty ships on rollers to the seaward side; the composition of this fleet is unknown but in view of the portage, unlikely to have included ships larger than a trireme. They attacked the Roman fleet about 15 miles off the city, both fleets stowing sails and deploying into line abreast. In a furious fight, the Roman flagship was boarded and their commander killed, then attacked by a second Tarentine ship, the flagship was captured. Most of the transports made sail and scattered, escaping while some of the Roman and allied ships were rammed and either sunk or forced ashore. This was the Roman's only naval defeat of the war and says much for the desperation with which the Tarentines fought.[14] The garrison held out and in 209 BC, Hannibal moved away. Covered by a fleet of thirty quinqueremes the Roman army advanced upon the city which was betrayed and surrendered. It would not be lost again but, could boast that it had been the only naval power to beat the Romans in all of its naval battles against them.

Egypt

Ptolemaic Egypt was one of the Successor Hellenistic powers that had fought each other to carve out empires from that of Alexander the Great in the fourth and third centuries BC. Much of the fighting had been at sea and had engendered a great naval arms race to out-build each other. They built more and ever larger ships, leading to the huge polyremes, Ptolemy eventually having thirty 'nines' and fourteen 'elevens' as well as larger and smaller ships; a 'sixteen' is the largest type to have taken part in a battle.[1] This arms race lasted from about 315 until about 250 BC. Ptolemy had finally managed to out-build his rivals and the wars had largely petered out by 275 BC, leaving three large powers (Macedon, Seleucia and Egypt) in an uneasy balance of power, each ready to exploit a weakness in the others, together with several smaller states. In 273 BC Ptolemy II established diplomatic relations and friendship with Rome and he and his successors maintained that relationship and neutrality throughout the wars in Greece, Asia Minor and the Levant in which the Romans became involved.

In the civil war between Caesar and Pompeius, the latter, when gathering his forces in Greece, called for and was sent fifty Egyptian warships for his fleet, the first break in their hitherto neutrality and one that would give Caesar an excuse to intervene in Egyptian affairs. These were placed under the command of his elder son Gnaeus.[2] They were sent to the Adriatic against Caesar's garrison and Dukat (Oricus) in Albania, blockading the port in increasingly difficult conditions between February and April 48 BC. The city is on a peninsula and Caesar's commander withdrew his ships into the protected lagoon behind it and blocked the harbour entrance with a blockship and a large merchantman upon which was erected an artillery tower. The Egyptians grappled and towed off the blockship and while making a diversionary attack on the tower and walls, used rollers to drag four biremes across the neck of the peninsula and launch them in the lagoon to attack the moored ships. Having also forced their way past the guardship, others of their fleet joined the attack, towing away four ships and burning the rest.[3] The Egyptian ships then continued to blockade the city, which nevertheless held out until Pompeius' defeat in August, after which they withdrew homeward.

Caesar's forces were trapped in Alexandria after his arrival there by the Egyptian's guard squadron of twenty-two ships, which was shortly joined by the fifty ships returning from Greece. These ships were related to have been quadriremes and quinqueremes,[4] the original guard squadron were only defined by Caesar as 'decked,' that is biremes or triremes. Of the remaining polyremes there is no mention, but their continued existence is perhaps confirmed by the appearance of 'tens' and 'sevens' within Antonius and Cleopatra's fleet at Actium. After the Alexandrine War had been won by Caesar, he left Cleopatra as pharaoh, with the remainder of the Egyptian fleet under her command and with a Roman garrison ashore. It was this fleet that she used to support Antonius in his eastern adventures that led to the civil war between them and Octavian. It was this fleet also that she took to Greece, sixty ships in strength and made up mostly of the quadriremes and quinqueremes that had faced Caesar, but with at least a few of the larger ships.

After Actium, Cleopatra and Antonius returned to Alexandria with their intact Egyptian fleet and a few of the Roman ships that had escaped with them. Early in 30 BC, Octavian arrived off Alexandria with his fleet, to blockade the city while his army advanced upon it. Antonius sent his ships and the Egyptian fleet out to oppose them, but had to watch

Philip V, King of Macedon (reigned 220 to 179 BC), in the Palazzo Massimo Museum, Rome (*Author's photograph*)

when the ships closed with Octavian's fleet and raised their oars in salute, changing sides. Egypt became a Roman province thereafter and its former fleet was to form the basis of the imperial *Classis Alexandrina*.

Macedon

By the start of the First Macedonian War in 215 BC, from being one of the leading naval powers, Macedonian naval power had become much reduced from the heady days of the arms race and polyremes. Philip V[5] nevertheless had expansionist plans towards the Adriatic and was anxious to isolate the Roman territory in Illyria. Thwarted by the arrival of a Roman fleet in his first attempt to seize a port on the Ionian Sea, he established communications with Hannibal, then in southern Italy and built up a fleet in Illyria of about 120 light warships. These were built by local shipwrights and were of their local type, *lemboi*, probably of both mono and bireme types. Philip advanced to attack Apollonia (near Fier, Albania) and Oricus, in 214 BC. A Roman fleet of fifty ships, mostly quinqueremes, with transports carrying extra troops and joined by ten allied Greek warships sailed for Apollonia. Completely outmatched, Philip's fleet was forced to retreat

up a nearby river until the big Roman ships could no longer follow. The Romans recaptured Oricus and drove Philip's forces back. Philip retired overland for the winter, burning all of his now useless ships as he went.

On his Aegean coast, Philip had a better fleet, but it was still outmatched by the Roman fleet of twenty-five ships, which had moved there in 210 BC and again in 209 BC, when it was joined by the allied fleets of Pergamum and Rhodes. Philip's fleet had three quadriremes and seven quinqueremes, plus three allied triremes and twenty or so lighter types. He took this fleet on a morale-boosting cruise around his garrisons but dared not seek an engagement with the allies. Acknowledging his position, Philip ordered the building of 100 new warships but the war ended in 205 BC, before many of these ships could be built.

Two years later Philip made use of his newly augmented fleet when, in alliance with Antiochus III they moved against Ptolemaic possessions in the Aegean and Levant respectively. Philip's fleet had now been built up to include a 'ten', several 'sevens' and 'eights' as well as quadriremes and quinqueremes, a total of over fifty large ships; he also hired pirate captains and their ships to augment his fleet. Philip attacked Pergamum, but was beaten by a joint Pergamene and Rhodian fleet off Chios in 201 BC.[6] Although he later beat the Pergamene fleet, the cumulative result of these actions was the loss of half of his fleet's strength.

Philip turned against Athens in 200 BC, as an ally of Rhodes and Pergamum; all were allies of Rome and then appealed to her, starting the Second Macedonian War (200–197 BC). Faced by a Roman fleet of thirty-eight ships, more than doubled by the addition of the fleets of Athens, Rhodes and Pergamum, Philip's fleet, reduced to at most twenty heavy ships, could only retreat to its home bases. They offered no opposition and after the battle of Cynoscephalae in 197 BC, which ended the war, one of the peace terms was that Philip had to surrender all but six ships of his navy. The surrendered ships were five 'light vessels' and Philip's flagship, a huge 'sixteen'.[7] This account does beg the question as to what happened to the rest of Philip's ships.

For the Third and final Macedonian War, Philip's son, Perseus had only a very small fleet, incapable of opposing the Roman and allied fleets. The fleet was not wholly inactive however and with about forty light ships it cruised off the coast of Asia Minor. The fleet managed to free and recover a convoy of Macedonian grain ships that had been captured by the Pergamenes; it also intercepted and captured a convoy near Chios, carrying cavalry horses for the allies. Despite these small successes, allied naval power remained unchallengeable and in June 168 BC the battle of Pydna ended the war; Perseus fled and Macedonia passed into the Roman domain, divided into three republics.

Pontus

Mithridates VI, king of Pontus on the Black Sea (reigned 120–63 BC) had expansionist ambitions and a fleet of nearly 300 ships. In 89 BC he attacked neighbouring Bythinia, an ally of Rome. He withdrew but the Bythinians counter-attacked whereupon Mithridates' forces overran Bythinia and much of the adjacent Roman province of Asia. With this, he sent his fleet into the Aegean in 88 BC, where they sacked Delos and went on to Rhodes. The king then sent his forces into Greece while his fleet commanded the northern Aegean. The dictator Sulla crossed to Epirus with five legions but no fleet and advanced to Athens.[8]

Although he drove Mithridates out of Greece, Sulla could not pursue him to Asia without naval support. He thus sent his deputy, L. Licinius Lucullus to gather warships, while Sulla marched his army to and through Thrace to the Dardanelles. Lucullus took three allied Rhodian ships from Athens to Alexandria where however, Ptolemy IX could not be persuaded to participate. He was more fortunate in gathering ships from Roman Asia, Cyrene and Macedonia and requisitioned crews from allied Greek states and islands. With this ad hoc fleet, Lucullus raided enemy coasts and covered Sulla's advance and crossing of the Hellespont. Sulla met Mithridates at Troy and a peace was signed in 85 BC, which included him

Prow of a Hellenistic warship on the grave stele of one Makartos from Delos, probably a lighter type, trireme or quadrireme. The epotis at the end of the oarbox can be made out. Early to mid-third century BC. Archaeological Museum, Dion, Greece. (*Author's photograph*)

surrendering seventy warships.

In 74 BC Nicomedes IV, king of Bythinia died and bequeathed his kingdom to Rome. Mithridates had recovered from the last war and rebuilt his fleet up to 400 ships. He could not countenance losing control over the Hellespont and so occupied Bythinia and as cover, again sent his fleet into the Aegean where, to divert opposition, they actively encouraged piracy all around Asia Minor. Marcus Antonius (senior) was sent to campaign against them while the consul G. Aurelius Cotta sailed for Bythinia with a fleet of sixty-eight warships and transports. His attempt at a land and sea campaign was foiled when he was beaten on land and at sea off Chalcedon by Mithridates' greater ship numbers, and his ships became confined to harbour. A raid by Mithridates' ships burned four Roman ships in harbour and captured most of the rest. The other consul, Lucullus, drove Mithridates' forces back, breaking the blockade and forcing him to withdraw his fleet.

In spring 73 BC Mithridates sent his fleet into the Aegean once more but was caught by a Roman fleet under Lucullus, who captured thirteen ships. Lucullus then caught another of Mithridates' squadrons at Lemnos and destroyed them on the beach. The survivors of

Mithridates' fleet were forced back into the Black Sea, where they suffered further losses from a storm. Lucullus' fleet also entered the Black Sea and supported the Roman advance along the north coast of Asia Minor until Mithridates was forced to flee to Armenia. Although great in numbers, the composition of Mithridates' fleet is not known, but its lack of prowess against a Roman fleet of sixty-four ships (assumed to have been mostly quinqueremes, their standard major warship type) plus some allied Rhodian ships, which, under a competent commander such as Lucullus, completely outmatched them, would suggest that they were predominantly smaller types.

Mithridates had managed to regain his kingdom by 66 BC but ran foul of Pompeius and his war against the pirates, those same whom he had encouraged and enlisted in his previous endeavours. Mithridates was again defeated and fled to the Crimea. Pontus, with Bythunia was constituted as a client kingdom and Pompeius left a strong squadron of warships and a garrison at Sinop (Sinope) backed by whatever of Mithridates' former ships had survived to serve the client king Pharnaces II. His successor, Polemo II, retired in AD 63, when Pontus became a province, the royal fleet being merged with the *Classis Pontica*.[9]

Seleucia

The other large power to emerge from Alexander's empire, Seleucia had originally included most of his Asian domains. In 185 BC it still included south-east Asia Minor, Syria and stretched across southern Iran although this was largely lost to the growing Parthian Empire in the following forty years.[10] Importantly it did include the Phoenician coast with all of its seaports, seamen, shipyards and ship-building tradition. It had been a major participant in the power struggles and still had a large navy which included some of the great polyremes.

With the defeat of Macedonia, Antiochus III saw an opportunity for expansion in Asia Minor and beyond. In 197 BC he took the important coastal city of Ephesus and moved across the Hellespont into Thrace. He used a fleet of forty heavy and sixty light warships to escort his troop convoy when he seized Volos (Demetrios) in Thessaly in 192 BC. He next advanced across Greece but in 191 BC, the Roman army returned to Greece and forced him back, before, with their allies, defeating him at Thermopylae, after which he withdrew to Asia. To prosecute the war there, Antiochus' powerful navy would have to be overcome. Antiochus' main base was Ephesus, where his admiral Polyxenidas had a fleet of seventy large, well-founded ships.

In their first encounter with the Romans and their allies, off Cape Corycus in September 191 BC, the Seleucid fleet was outnumbered and worsted, losing ten ships destroyed and thirteen captured, before withdrawing to base. The allied fleet was then strengthened by the arrival of the Rhodian fleet, bringing its total to 127 ships and leaving Polyxenidas unable to challenge.

Over the winter of 191–190 BC, Polyxenidas' fleet was built up to nearly ninety ships, including two 'sevens' and three 'sixes'. The opposing Roman fleet was of quinqueremes with some triremes, the Rhodians preferred lighter triremes and quadriremes, while the Pergamene fleet could boast a few 'sixes' and 'sevens'. Additionally however, Hannibal, who had fled Carthage and taken service with Antiochus, was mustering another fleet in Syrian ports and when it sailed he had forty-seven ships, namely three 'sevens', four 'sixes' ten triremes and thirty quinqueremes and quadriremes from Phoenicia. He was defeated and although only losing one ship, a 'seven' and a dozen damaged, withdrew and this fleet did not again see action.

Polyxenidas had eighty-nine ships, including three sixes at Ephesus. After some manoeuvring, his fleet met the allies at the battle of Myonnesus in October 190 BC and was soundly beaten, losing twenty-nine ships captured and thirteen destroyed. Antiochus' naval power had been broken, despite still having the thirty-seven ships of Hannibal's fleet and forty-seven survivors of Polyxenidas' fleet. In the peace that ended the war, Antiochus surrendered all but ten of his ships. Seleucia continued to exist but never again as a naval power, until finally, Pompeius in 63 BC annexed Syria as a province.

In the north

Roman expansion in the second and first centuries BC had brought their naval forces in to the Atlantic Ocean and around the coasts of Spain and Portugal without meeting any naval opposition. In the mid first century BC however, Caesar's campaigns in Gaul brought him, in 56 BC, to the coasts of southern Brittany, home of the Celtic Veneti, a seafaring race who had evolved sturdy, seagoing craft suited to their often stormy coasts and who voyaged and traded widely on the French Atlantic coast and on both sides of the English Channel. Caesar describes their ships: 'The Gaul's own ships were built and rigged in a different way from ours. Their keels were somewhat flatter, so they could cope more easily with the shoals and shallow water when the tide was ebbing; their prows were unusually high and so were their sterns, designed to stand up to great waves and violent storms. The hulls were made entirely of oak to endure any violent shock or impact; the cross-beams, of timbers a foot thick, were fastened with iron bolts as thick as a man's thumb; the anchors were held firm with iron chains instead of ropes. They used sails made of hides or soft leather, either because flax was scarce and they did not know how to use it or, more probably because they thought that with cloth sails they would not be able to withstand the force of the violent Atlantic gales, or steer such heavy ships.'[1]

To oppose these ships, Caesar brought warships from the Mediterranean and built more in Roman Atlantic ports. The Gauls amassed their own fleet, some 220 in strength for the final, showdown battle, a contest between two totally different shipbuilding traditions and philosophies. The Gaulish ships were impervious to the Roman rams and they were too high out of the water to permit easy boarding, even turrets on their decks would not give the Romans sufficient height advantage. The Roman ships, powered by oars, were faster and more manoeuvrable, while their opponents were powered solely by sail. The Romans had however, equipped themselves with long poles, tipped with scythe-like blades and as they came in close alongside the Gauls, they seized and cut their rigging, causing the yards and sails to collapse, whereupon several Roman ships could close and board. Seeing this, the Gauls tried to sail off but the wind dropped and they were stopped and at the mercy of the Romans.[2]

With the acquisition of Gaul and Britain, Celtic shipbuilding and seafaring became absorbed into the Roman Empire and after Caesar nothing is heard of any maritime opposition to Roman naval power, with the exception of the Bructeri on the River Ems in 12 or 11 BC, who sought to oppose the Romans by using their boats. There followed a naval battle, which the Romans won, but no details of the engagement survive. There was no naval opposition during the invasion of Britain. In the late first century AD, the writer Tacitus was with Roman forces in Britain and described the ships of the Suiones of southern Scandinavia as follows: 'The shape of their ships differs from the normal in having a prow at each end, so that they are always facing the right way to put into shore. They do not propel them with sails, nor do they fasten a row of oars to the side. The rowlocks are moveable, as one finds them on some river craft and can be reversed, as circumstances require, for rowing in either direction.'[3]

Notable is the lack of any form of sailing rig, indeed there is no evidence that sailing ships appeared in the (non-Roman) north before as late as the seventh century AD.[4] However, tribes inhabiting the North Sea coasts will have become familiar with Celtic trading vessels far earlier and would see Roman ships with sails from Caesar's time onward. The copying and adoption of such an obvious and in principal, simple device by those tribes from the first century AD could be assumed.[5] In support of this, Tacitus mentions that in the final foray of the rebel Civilis' fleet, made up of captured Roman warships, both monoreme and bireme, plus large numbers of native craft and other captured ships, that they were rigged with improvised sails.[6] Also absent from barbarian ships was any form of ram, a feature which they never adopted. The rams mounted on Roman ships, together with artillery, ensured their continued advantage in a sea battle.

It is from the late second century that the situation changed with the start of piracy and raiding by the Germanic tribes of the North Sea littoral. Although not constituted as a 'navy' as such, groups of tribesmen from all along the coast gathered in their boats to raid and steal along Roman coasts and up rivers. It is from this time that Roman coastal towns began to receive defensive walls and forts started to be built at salient points on the coasts. Whether or not they had sails, or relied solely on oars, barbarian seamanship improved and they were able to cross the southern North Sea by the early third century AD. Units of the British and German fleets were moved to forward bases in Kent and the Thames and Rhine estuaries, to combat this piracy and raiding. Continuing prosperity in the affected provinces speaks as to the success of the defensive measures, which were to be progressively augmented through the third century AD, as well as the comparatively low level of barbarian naval activity.

The boats used by these early raiders were open, clinker-built double-enders (as described by Tacitus) and rowed by up to fifteen men per side. North Germanic plank-built craft date from at least 350 BC, the date of the earliest example of one so far found, some 52½ feet in length (25.4 m) with thwarts for twenty paddlers.[7] The remnants of a small boat from Halsenoy in Norway and dated to the first century AD, included rowlocks of a fixed, spur type and proves that from that time at least, barbarian craft could be rowed rather than paddled.

During the third and fourth centuries AD, the Germanic tribes of the North Sea littoral appear to have progressively formed themselves into larger confederations.[8] Better organisations meant that greater numbers of ships could be built, manned and concentrated for increasingly frequent raiding by larger numbers of ships upon Roman territory. Barbarian seafaring capability improved and their vessels grew in size, the mid-fourth century oak ship recovered at Nydam in southern Jutland being 79 feet (24 m) in length, with fifteen thwarts to seat rowers. With thirty rowers and a dozen or so other men on board, a raid by only a dozen such ships represented a force of about 500 men, a serious threat to Roman coasts. From the early third century AD the Roman defensive system was greatly increased by the building of strong fortresses, the so-called 'Saxon Shore Forts' along Channel coasts, at each of which, apart from the garrison, was stationed a force of warships. They were supported by a string of smaller forts and watch towers, to provide a warning system.[9]

Overall, these strategies were successful in minimising or even thwarting barbarian naval activity, at least while the system was able to be kept fully manned and maintained. Although the barbarian craft were evolved to be good sea boats, able to cope well with the coastal voyaging and short sea crossings, technologically, Roman warships remained superior, their opponents, from the limited evidence available, showing no technical evolution into dedicated warships. Their effectiveness relied upon numbers and in having the initiative granted to the unexpected attacker. Seaborne raiding increased in size and frequency from about AD 230 and would eventually become endemic.

Whatever motivated the barbarians to adopt raiding and plundering almost as a national pastime, a number of factors within the Roman world encouraged them. First, they were aided by the period of crisis within the Roman Empire from the murder of Caracalla in AD 217 to the triumph of Diocletian in AD 285. During that period, Roman forces were often focused on internecine struggles at a time of increasing barbarian unity and strength, which created opportunities for them and left a perceived, if not an actual weakness in Roman military strength which they could exploit.

The other factor which had a profound effect on the north coasts was the great inundations of the mid-third century AD, caused by a rise in the level of the North Sea and subsidence of the lands affected. This created the Zuider Zee/Ijssel Meer, the Dutch islands and the Wash and submerged much of the coastal areas of Belgium and Flanders, reducing vast areas of the remaining land into sodden, salty marshes. The consequent disruption and ruining of agriculture forced many barbarians to turn to raiding for survival and at the same time, rendered untenable and de-populated swathes of Roman coastal lands between the mouth of the Rhine and Cap Gris Nez for many miles inland.[10] It also made much of the area indefensible and as a result the provinces in Belgium and northern Gaul suffered destructive raids and damaged economies. The British south coast was largely unaffected by the inundations and although suffering raids, they appear to have been on a much smaller scale there. Nevertheless, seaborne raiding continued and grew until the English Channel and North Sea coasts of the Empire were infested with barbarian craft.

In AD 285 Carausius was appointed commander of Roman forces in the Channel.[11] He rebuilt the fleet and took the offensive, intercepting barbarian craft, usually on their way home when their presence, strength and location were known. The barbarians were driven out but the emperor, possibly with some justification, accused Carausius of embezzling the recovered loot; in AD 286, Carausius declared himself emperor. Throughout his secession and that of Allectus, his successor, defeated by Constantius in AD 296, the Roman fleets remained strong and barbarian raiding was severely curtailed.

The barbarians were not quiescent for long and raids took place along the Atlantic coasts, as far as Spain in AD 313. Whether they used their traditional craft, captured Romano-Gallic ships or a mix of the two is not known, but the voyage opened a new chapter in their activities and demonstrated their vastly improved seafaring ability. After AD 350 this ability became manifest when pirate raids became more frequent and the virtual collapse of the lower Rhine frontier left the defences hard pressed; the myriad inlets, channels and islets left by the inundations provided havens for the barbarian craft and also access to areas formerly too far inland and unreachable by boat.[12]

By the fourth century AD Scottish pirates appeared, operating from Ireland and southwest Scotland, as well as Pictish equivalents on the east coast of Britain.[13] In AD 367, a barbarian conspiracy among the tribes of Scotland and coastal Germany launched co-ordinated attacks on Britain and northern Gaul which overwhelmed the defences and largely overran the countries causing immense damage. The situation was temporarily restored by the future emperor Theodosius, but support in Britain for the usurper Magnus Maximus in AD 388 drew away many of the armed forces which had been left to oppose the barbarians to Gaul and permitted a resumption of their activities and attempts to seize and settle land within the Empire.

In AD 396 Roman naval forces returned to eject the invaders but any security offered by them was illusory, the barbarians were too numerous and roamed at will across the seas. Although much of the shore defence system was still manned in the early fifth century AD, Roman naval activity was nominal and concerned with keeping open the cross-Channel lines of communication. Any semblance of a fleet had gone and with the progressive takeover of Britain and parts of northern Gaul by barbarians, came the abandonment of the forts during the fifth century AD and the distinction between barbarian and Roman ships disappeared, as indeed, did Roman ships.

In the south-east

Roman power was introduced to the Black Sea by the wars against Mithridates of Pontus in the first century BC. As a result the Romans found themselves in possession of the south coast of that sea. To the north, in the Crimea and the lands around the Sea of Azov (Maeotis Palus) was the semi-Hellenistic kingdom of the Bosporus (Regnum Bosporum); along the west and north-west coasts were long-established Greek trading cities, none of which offered any naval opposition to the Romans. After defeating Mithridates for the last time in 66 BC, Pompeius marched, accompanied by his fleet through Armenia and into the Caucasus, modern Georgia. A sweep of the east coast of the Black Sea by his fleet confirmed that it had become a Roman lake. Local states policed their own waters with light craft but had no warships as such. In AD 46, Claudius annexed Thrace making the whole Black Sea coast from the Danube Delta nearly to Poti in Georgia either Roman or controlled by them through client rulers. The peace of the Black Sea was disturbed during the civil war of AD 69; warships had been withdrawn to the Bosporus in support of Vespasian and in their absence there was an uprising in Pontus (annexed in AD 63) led by the former commander of their Royal fleet. The rebels seized Trabzon and to whatever warships they could find added hastily built boats. These were broad-beamed double-ended boats, described as narrow above the waterline, presumably therefore having tumble-home and capable of being rowed in either direction, i.e. double-ended. Additionally freeboard could be increased by adding washboards the extent that the boats could actually be completely enclosed. These 'rebels' were in fact little more than pirates and started to roam the sea accordingly. Vespasian quickly sent a force of warships and troops who drove the rebels to their ships and who fled to Georgia; there the approach of the Roman fleet persuaded the local ruler to surrender them to the Romans.[14] Despite the Roman's overall hegemony, there remained large stretches of the eastern, Caucasus coastline from which the occasional marauder ventured and one or two even penetrated into the Aegean; there was a low level of coastal piracy in the first and second centuries AD, the pirates using small boats called a *camara*, each with twenty-five or

thirty men and capable of being lifted ashore and hidden.[15]

The situation was to change in the third century AD which was a period of crisis when the fleets protecting the Black Sea suffered neglect, loss of morale and a deterioration that left them much weakened. Similar effects upon the Praetorian fleets of the Mediterranean meant the loss of their backing of trained men, ships and equipment. To the north, Germanic peoples had been migrating southward towards the lower Danube and Ukraine. Gradually the Greek cities of the north Black Sea were overrun. In AD 238 the Goths crossed the lower Danube, raiding into Thrace and sacking the *Classis Moesica* base at Istrus in the delta; for this crossing they needed only such boats and rafts as they could obtain, but in substantial numbers.

By AD 251 the Goths and Scythians had amassed a fleet of some 500 craft of all types on the north Black Sea coast. This fleet sailed and managed to slip through the Bosporus and into the Aegean unchallenged. They attacked Macedon, Greece, Athens, Sparta and Corinth before being caught and stopped by Roman fleets.[16] The Goths had by now all but taken over the Bosporan kingdom and from them and the peoples of the northern Black Sea, were learning how to build and use ships. Although they only had small, flat bottomed boats, the barbarians sailed in large numbers, carrying hundreds of men and descended upon undefended, unsuspecting towns that had lived in peace and security for over two centuries. As in the North, the sheer numbers and frequency of barbarian forays and the comparative numerical weakness of Roman fleets, made them difficult to intercept. Roman ships remained greatly superior and could destroy them if they caught them, raiding fleets were beaten in battles off Rhodes, Crete, even Cyprus but many more must have made it home, with their loot.

In AD 254 raiders used seized Bosporan ships with captive crews, indicating perhaps that their own, native craft remained small and that they had not yet acquired the ability to produce or use anything more sophisticated. Two years later, the Goths again raided the north Anatolian coast, reverting to the use of their small, open boats. In AD 259 another mass raid by up to 500 barbarian craft penetrated into the Aegean, attacking Athens, rounding Greece to attack Epirus on the Ionian Sea and returning to destroy the Temple of Artemis at Ephesus before withdrawing. Of Roman naval opposition there is no mention. An attempt to repeat the feat in AD 269 saw the Goth fleet land and actually attempt a siege of Thessaloniki. The following year, the survivors were driven off and after desultory attacks on Rhodes, Crete and Cyprus, they were defeated and totally destroyed by the *Classis Alexandrina*.[17]

The situation again changed in AD 330 when the Constantine moved the Empire's capital to his new city of Constantinople, built atop the old Greek town of Byzantium on the Bosporus. The greatly increased Roman military strength in the area consequent upon this deterred further attempts at penetration by the barbarians into the Aegean. In fact their naval activity had all but ceased by that time as the focus of Goth attention had moved landward to the lower Danube area. The overrunning of the north Black Sea area by the non-maritime Huns in the late fourth century AD ended barbarian naval activity there.

The Vandals

The western Mediterranean, remote as it was from the areas of barbarian activity, enjoyed relative freedom from their depredations. That freedom was interrupted in AD 258; a large number of Frankish tribesmen had crossed the Rhine and avoiding the frontier defences, they then proceeded to pillage their way across Gaul and Spain to Tarragona which they sacked. That they were able to do this was a consequence of the defensive system which concentrated military strength on the borders; once past them, there was little to interfere with the Frank's progress. The nearest legions were stationed in north-west Spain and North Africa.[18] In Spain, the Franks seized a number of merchant ships (presumably with their sailors) and sailed for Africa[19] where, after some raiding, they were repulsed. They may have been operating as pirates from Spain for some years in fact and

their ultimate fate is unknown.[20]

Twenty years later, an experiment to resettle large number of captured Franks in Pontus went disastrously wrong. As has been seen, these coasts were already suffering barbarian raids and in AD 279, taking advantage of the distraction caused by a civil war, the Franks rebelled, seized as many ships as they could and set off on an epic voyage of piracy. They sailed into the Aegean, attacked Cyrene (eastern Libya) then Sicily, where they sacked an unsuspecting Syracuse. From there, they crossed to Tunisia, where they were driven off by local forces. The voyage continued on, into the Atlantic and they rounded the coasts of Spain and Gaul to return to their homeland.[21] This was an amazing feat of seamanship and significant in that at no time were they challenged by Roman naval forces, indicating perhaps how far the latter had deteriorated. It is not known how many Franks set out, or how many ships they had, or ultimately how many survived the voyage.

In the winter of AD 406 the Rhine froze solid, preventing Roman river patrols and permitting a mass crossing by tens of thousands of Vandals, Alans and Suevi. They rampaged across Gaul and in AD 409, crossed into Spain. The Vandals, doubtless joined by other tribesmen and various malcontents from Gaul and Spain, ranged over and plundered Spain for some twenty years. It seems likely that opposition built and they moved to southern Spain, there, they started to acquire some ships and raid the African coast and Balearic Islands. Finally in AD 429 they crossed to Africa en masse[22] and for the next ten years, gradually made their disruptive way eastward across North Africa. They were at last opposed by a Roman army of between ten and twenty thousand men at Carthage, commanded by the count of Africa, Boniface. Incredibly, Boniface crossed to Italy with the army seeking power, leaving Carthage virtually undefended.[23] The Vandals took Carthage in AD 439, completeing their occupation of northern Tunisia and setting up their kingdom there. With Carthage, they acquired a fleet, shipyards and the people to build ships, as well as experienced crews.[24]

For the first time in over six centuries, there was a navy in the central Mediterranean other than Roman. The Vandals proceeded to build up a fleet of some 120 warships and with them, went on to occupy much of Sicily, Sardinia and Corsica by AD 440. The eastern emperor sent a hastily formed naval expedition against them which failed. Other diversions meant that the Empire could not focus on the Vandals who, in AD 455 sailed to Ostia and attacked and sacked Rome.[25] A Roman fleet from Italy did take the offensive and beat the Vandal fleet off Corsica in AD 456. The Vandal advance in Sicily was halted and their fleet again beaten off Agrigento. In the following year (AD 457) the Vandal fleet was ejected from Ostia and the mouth of the Tiber.

In AD 460 the western emperor Majorian (reigned AD 457–461) was fitting out a fleet in Cartagena of some 300 ships but the Vandals learned of it and in a surprise attack, destroyed the fleet before it was ready and able to sail. The Vandals went on to raid the Peloponnese in AD 467, provoking the eastern emperor Leo (reigned AD 457–474) to action. The eastern half of the Empire had a large fleet and this escorted a huge army intended to eliminate the Vandal kingdom. The western emperor Anthemius (reigned AD 467–472) sent forces and his fleet, which had in the meantime, driven the Vandals from Sardinia. The whole force was commanded by Leo's brother-in-law Basiliscus. To attack Carthage he chose to make his landing in the Bay of Tunis against the west face of Cape Bon (Hermaeum) a rocky, lee shore. In both previous Roman invasions of Carthage, in the Second and Third Punic Wars, they had landed and made their base to the west of Carthage, and thus always had the weather gauge in subsequent operations against the city. Instead of attacking at once, when his forces would have swept all before them, he delayed. This gave the Vandals time to prepare and five days later the wind went round to the west, holding the Roman fleet against the shore and enabling the Vandal fleet, with the wind behind them, to attack with fireships and the ram. The closely packed and anchored Roman ships had no room to manoeuvre and half of their ships were destroyed, the rest escaped to Sicily, abandoning the campaign.[26] This was the last operation of a fleet of the Western Roman Empire, the fleet of the eastern Empire did continue in existence to be what we now, for convenience, refer to as Byzantine.

APPENDIX MODELLING NOTES

MODELLING ROMAN SHIPS

THE FIELD OF MODELLING Roman warships is a wide one and ideal for the modeller looking for something a little different yet which allows scope for interpretation and the imagination. The modeller has been ill served in the matter of proprietary kits of Roman ships. Apart from the Imai/Academy plastic kit of a 'Roman Warship' and a more recent Russian plastic kit, there was a wooden kit of 'Caesar's Galley' from one of the European kitmakers some time ago and some miniature metal models by Skytrex for wargaming. In all of these examples the interpretations represented by them are, of course, subjective and the comparative paucity of kits perhaps represents the manufacturers' evaluation that Roman shipping has not been of much interest in the modelling fraternity.

The very vagueness in the surviving accounts and iconography can in fact allow scope for the modeller's imagination, bearing in mind always the limits imposed by the materials, technology and knowledge of the time. There are other parameters that must be considered in the modeller's preparation for building a miniature Roman warship. Most importantly, having decided on the type, the number and accommodation of the rowing crew, the 'engine' of the ship, must be so arranged to ensure that each of the nominal rowers at each level has room in which to operate, this is especially so in a bireme or trireme. As has been seen from the descriptions of the ships, the size and shape of the rowing compartment will dictate the form of the whole ship, as will the adoption of correct length to beam and draft to hull height proportions. All of this nevertheless allows a lot of scope and permits considerable leeway in interpretation, allowing one produce a model which can with justification, be said, in the modeller's opinion, to be of a Roman ship.

Research

Before embarking upon the model, a plan will be required; although there are a few drawings in books, the modeller is left with little option but to draw his own. This in turn leads to the necessity for research, which is concentrated in three areas. There are the surviving writings of ancient authors, contemporary with these ships; unfortunately this literature rarely gives any clear description of the ships, with which their original readers were in any event, familiar. The odd valuable detail can be gleaned; for example, Caesar moored his ships offshore in 54 BC, from which we can gather that they were equipped with anchors. (Caesar, *De Bello Gallico*, IV.29; also Livy, *History of Rome*, XXV.25)

An obviously fruitful source of information on the appearance of Roman warships is iconography. This can be found as wall paintings, such as those recovered from Pompeii, Herculaneum and other sites around the Bay of Naples. Although impressionistic in style, these paintings have the advantage of being of the ships that the artist could actually see, just across the bay at Misenum. Although several mosaics show merchant ships, any with depictions of ships that could be warships are very rare and tend to be very stylised; but again, the odd useful detail can be discerned. The columns of Trajan and Marcus Aurelius in Rome have ships with a great deal of detail but suffer from the common convention of showing men aboard them oversize and out of scale; they are also distorted by the shape and size of the stone panels into which they have to fit. Other statuary, especially grave stelae, also show ships or parts of ships. Many Roman coins had ships or ship's prows on them, especially in the Republic.

In all of these cases, correlation of a detail from more than one such source will corroborate the accuracy of it and add authenticity to the model. The whole field can be engrossing and its study will lead to the building up of a body of information for the modeller which, when applied, will produce a more authentic and satisfying model.

Scales

The choice of scale for a model really depends on how big you wish the finished model to be. There is a choice between imperial or metric, but it matters not as long as they are not mixed on the same model. For example, where the scale is 1:300, that is, the drawing is one three-hundredth the size of the real thing, whichever system is used, multiplying the measurement by three hundred should yield a full-size ship that should actually work in real life, or at least very much look as though it would. The author prefers imperial, having been brought up with it, and feels that it is more fitting, since 'imperial' is very close to the sizes that the Romans themselves used for the originals. One point worth mentioning is that if you intend to make more than one model, it is worth considering using the same scale for them, which gives a ready comparison between them and results in a satisfying display.

Using a small scale, such as 1:300th (25 feet to the inch) has advantages; a stock of materials goes a long way and you need only a small space to display (or to store) a lot of models, but most importantly, much of the detail of a ship can be left out. Remember, the larger the scale of the model, the bigger it is and more of the smaller fittings and details will become visible to the naked eye and will thus have to be included in the model. Just imagine a very small scale drawing at twice or more times as big and you will see how the 'gaps' grow and that more conjecture will be required by the modeller to fill them in and thus less accuracy can be attested for the model. Conversely, unless more fine detail is included, the resultant larger model will look sparse and quite unsatisfactory. This, of course, goes hand in hand with the other big problem, which, as has been seen, is that few particulars survive as to the nature of those ship fittings and details.

None of this is to say that one should not undertake models at larger scales, to give variety and enable different techniques to be explored and for more detail to be included, especially if one wishes to experiment with, for example, reproducing rowing systems, or even making a working model.

Whichever scale is chosen, it is important to have a ruler or scale which reflects it and from which parts can be measured and worked. As an example, in the 1:300th scale, a regular school-type ruler, with a section in inches and divided into tenths of an inch, yields a useful two feet six inches for each such subdivision and so on. Metric rulers at various scales are also available and similar in concept for our uses. If modelling in any constant scale, it is essential to invest in an appropriate rule, or even to make one.

Imperial scales are simple to use as they, of course, divide exactly and infinitely and a standard rule can be used with all of its subdivisions: one inch is 1/12th scale, a half inch is 1/24th scale, a quarter-inch is 1/48th scale and so on. Conversely at 1/8th-inch scale (1:96), 1/8th inch equals one foot, so 1/16th represents six inches, 1/32nd is three inches and 1/64th is only one and a half inches; and that really is about as small as one need go. Another point to bear in mind is the strictly organic nature of these ships; the most important feature was that they had to be built around the human frame that was their motive power and so ultimately the scale is that of the crew.

When deciding on a scale you should consider how big you might want a finished model to be. For example, the longest Roman ship recorded in the literary record was a special ship built for the emperor Gaius to transport an obelisk from Egypt to Rome: it was 320 feet long by 65 feet beam (106 by 20 m). It would yield a model at 1:300th scale that would be 12.375 inches (315 mm) long and at 1:192 (1/16-inch scale) still only 14.575 inches (307 mm) in length.

Tools

The tool kit need not be large or exotic and is best built up gradually, starting with a few basic items and adding further tools as requirements demand. There is also of course the occasional 'must have' item that is spotted from time to time. Quality does pay and a good tool can serve well and last a lifetime (of the modeller that is).

First and foremost is the need for somewhere to operate the model shipyard and a sturdy baseboard, big enough to accommodate the drawing from which you are working and the work area itself, as well as the 'toolbox'. This can be a table, desk or whatever set aside for the purpose or a suitable board, perhaps with baize backing, able to be put on another table and which can be cleared away when not in use.

A cutting mat, either a piece of softish wood or one of the specialised types available is well worth having to avoid damage to the baseboard.

A reasonable tool kit could comprise:

> cutting board
> modelling knife and blades
> fine pliers

small side-cut wire cutters
tweezers
miniature Archimedes' drill
small palm drill
fine drills
pin vice
sharp needles
scales to suit
steel rule
steel square
paint brushes
needle or mousetail files
small sanding block
small gouges and routers
fine sandpaper

Commenting on the above list and starting with the knives; for carving the author prefers an X-Acto no. 5 knife handle as the butt end fits nicely in the palm, leaving thumb and fingers free to guide it; it is paired with Swann

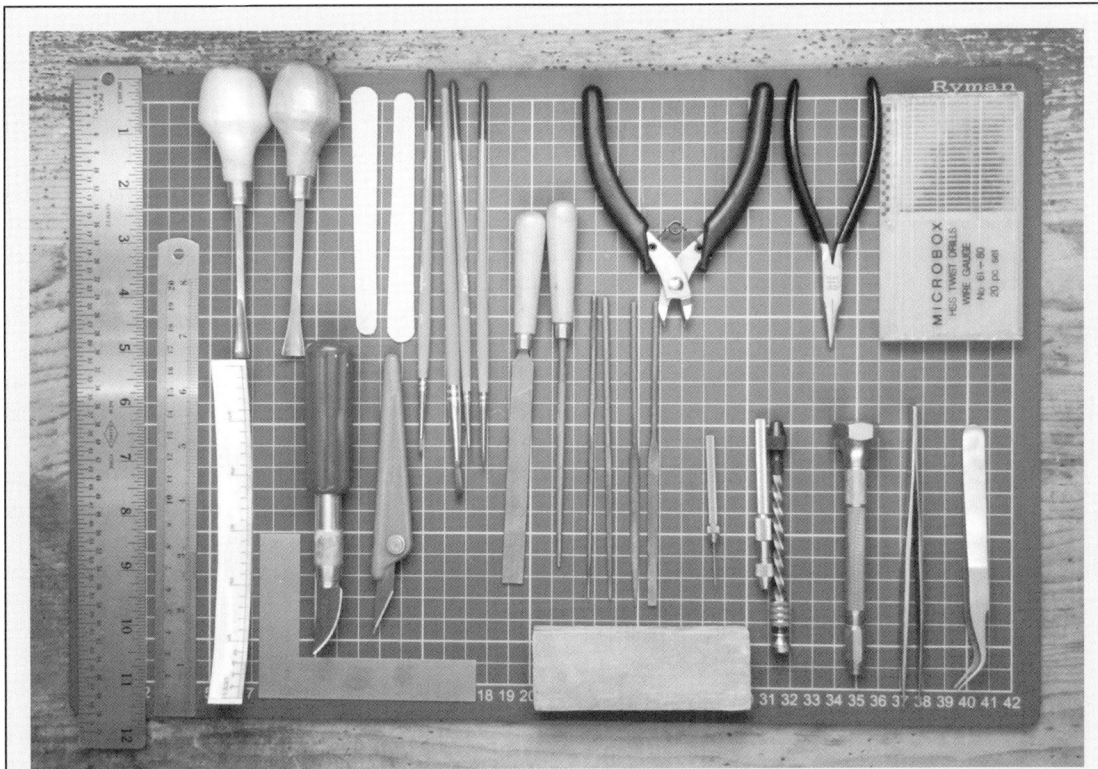

The tool kit. A selection of basic modelling tools as described in the text, set upon a cutting mat.

Morton no. 2 blades (the curved one), which being thinner than the X-Acto one, needs a sliver of plastic card inserted with it for the handle to grip well. For cutting straight edges, the no. 1 straight blade is preferred. These are only the author's personal preferences and you will quickly discover when working, methods and tools which may well suit you better. Additionally, chisel-edged blades will have their uses and very small woodcarver's round gouges and routers are good for hollowing out hull shells. In all cases, use only very sharp blades; it is the blunt one that slips or mis-cuts and causes accidents or inaccurate work. It should also go without saying that you should always cut away from yourself.

The smallest pair of pliers, with fine-pointed, smooth-faced jaws are essential for a multitude of tasks, especially when handling wire, as are the small wire cutters; ensuring that the latter are of the side-cutting type will give a square end to the wire and even to cutting or trimming of tiny wooden parts.

A selection of fine, wire-gauge drills will be useful and a most useful tool for all manner of marking out and making of tiny holes is a sharp, fine needle, held in a small pin vice.

Paint brushes; good quality does pay off in better performance and less finishes marred by the odd stray hair that has come adrift. Artist's quality brushes from size two-O up to about four, will suffice for miniatures.

For cutting straight pieces, a steel rule is essential; wooden ones are fine but, no matter how careful you are, it will soon develop 'moth-eaten' edges.

For abrasion, a set of very small files of varying cross-sections and a smallish block of wood with a level surface to which is glued a piece of fine sandpaper should handle most needs.

The above is not intended as an exhaustive list but at least it will yield a sufficient toolkit to make a good start on miniatures of Roman warships. Obviously it reflects the author's own preferences and usage and you will quickly develop your own favourite tools and methods for their use.

Finally, something in which to keep these collected treasures is more than desirable. There are any number of suitable plastic cabinets with a variety of drawers in all sorts of configurations available and most suitable for holding your variety of tools, materials and the collection of 'bits' that will inevitably build up. Small wooden chests of drawers and such like can be used, or you may even wish to build your own bespoke version; it is basically a box of convenient size to include drawers and compartments to house tools, paints, materials and the odds and bits that go to make up your model shipyard.

Materials

Wood

Various woods can used, chosen according to its properties to suit a particular requirement, the most important being balsa and lime. Balsa, the modeller's staple for many decades, is light, soft and very easy to carve; it is also easy to mark, can become 'fluffy' when sanded and does not easily give a good, painted finish. For making hull plugs, parts that are going to be hidden or even mock-ups as trial pieces however, it cannot be beaten. Lime is the wood carver's timber *par excellence*. It is expensive, but a little goes a long way and it is well worth seeking out a good piece of well-seasoned, preferably English lime from a specialist timber merchant. It has no equal for miniature work, being relatively easy to work, straight-grained and knot-free. It can be worked to incredibly fine definition, with a finish so smooth that it requires virtually no sanding; it takes paint well and can be fashioned into practically any form needed and into tiny pieces, strips and shavings. Easily the most useful all-round wood and a must-have. Other specialist woods can also have their uses; basswood or American lime is (not so good but) easy to work, straight-grained and can be cut to fine tolerances but leaves a few more 'whiskers' than lime. Well-seasoned pear is harder but gives a beautiful, silky finish. What is desired is a straight, close, knot-free grain in a wood that can be worked with the modelling knife, so oak for instance, although beautiful, is hard to work and not really suitable.

Dowel in 1/8th or 1/16th inch (and even

1/32nd if you can find it) sizes can be whittled and sanded for masts and spars.

Paper

Various thicknesses of paper and thin card can be used to make sails, awnings, hatch covers and bulwarks or screens. Avoid any with watermarks which may show through, even if you think that they will be invisible.

Plastic card

This also comes in varying thicknesses, has some uses, but really only for parts that are going to be painted as otherwise, the stark whiteness of the plastic will look wrong.

Wire

Wire is widely used for oars, ropes, rigging and a multitude of small fittings. A card of electrical fuse wire gives a selection of diameters and copper wire in a variety of useful diameters, right down to incredibly fine, can be recovered from bits of electronic and electrical cable; telephone wire is particularly useful; seek and ye shall find and put it down to recycling. All are soft and easily cut by the cutters and the variety will enable an appropriate diameter of wire to be selected for each application.

Fillers

Fillers will be needed from time to time and a tube of fine surface filler from the DIY shop will provide a finish that is fine enough for our purposes and require little extra finishing. Plastic wood is an alternative but has a coarser texture and needs more finishing. In very small scales, the use of fine surface filler has the advantage that surplus can be carefully scraped with the edge of a knife without spoiling the surface. For parts that are to be cast, a small tub of car body filler can be utilised.

Metal foil

The metal foil still to be found around the necks of wine bottles should be carefully removed, smoothed out and hoarded, especially as it seems to be becoming more rare. It can be used for any part that is of a curved, round or awkward shape, such as a rounded bulwark, the flukes of an anchor, the 'skirt' of a ram; cut into strips it can make straps on anchor stocks, brackets, small metal fittings and such like. The foil's very flexibility and the fact that once formed into the desired shape, it will retain that shape makes its uses legion.

Adhesives

Cyanoacrilate, 'superglue' and white PVA glue generally satisfy the need for fast-acting and slow-acting glues respectively. The added advantage of PVA is that it can be painted on to provide a protective coating but when dry, is colourless and completely matt, an advantage on wood that is to be left in its natural state.

Finishing

For finishing, a coat of knotting applied to balsa will firstly toughen the outer surfaces to better withstand handling and give a surface capable of being sanded to a finer finish for final filling and painting. Paper and card parts, such as sails, if painted before fitting to the model in ordinary water colours in muted tones, will yield a finish which looks, in the Author's opinion at least, just right for an ancient ship. There again, should you chose a more gaudy finish, who is to say that you are wrong, certainly ancient statues were painted in vivid colours. For other paint coverage, modeller's matt enamels or acrylics should fulfil all needs. The use of watercolours can particularly give a pleasingly soft and faded appearance that looks in keeping with the tones of surviving Roman wall paintings; this effect may look correct to our eyes, but we have never seen what those paintings looked like when new, perhaps they were bright and vivid like the colours still to be seen in even more ancient Egyptian tomb paintings. Perhaps the Romans did paint their ships in vivid colours, but until one is found, your own preference is valid.

MODELLING TECHNIQUES

Hulls

There are several methods for making hulls for the models. First, a solid lump of wood can be carved to the shape of the outside of the hull; the block is cut to slightly larger than the overall dimensions of the finished hull. For either full-hull or waterline models, start by levelling or squaring the face that will become the bottom. The upper deck sheerline can then be marked up from the bottom; it is much simpler to transfer the sheerline from the plan to the flat surfaces of the sides of the block (it should be marked on both sides) than to do so once it has been carved to the hull shape. Deck sheer on ancient warships is only notable at bow and stern, the whole centre portion of the hull, accommodating the rowing crew, was virtually without sheer, so that the rowers and their oars, were all at the same height above water. The deck sheer can then be carved, following which, the centre and keel line is marked from stem to stern (or vice versa). The hull sides can then be carved to their final shape and sanded. Prepare from thin strip, stem, keel and sternpost pieces; make a shallow cut along the stem, keel and sternpost lines of the hull underside and once you are satisfied with its accuracy and straightness, widen it with the back of the blade to accept the keel pieces.

A second method is to carve and finish the hull as above and then to 'plank' it with thin strips of wood, cut from lime shavings. It is fiddly and only worth doing of course, if the hull is to be left in a natural wood finish, when the result can be very pleasing in resembling a 'properly' planked hull.

Third, for open, i.e. undecked, hulls, a block is carved to the finished shape of the hull but wider along the centreline by the thickness of the keel. The hull is then carefully cut in two along the centre (keel) line and reduced to the beam minus the keel thickness, as it will be added later. Either one, for a half-model, or both halves are then hollowed out with gouges so that the resultant shell halves are about 1/16th inch (1.5 mm) in thickness all over. Having of course, prepared the keel, each half is then glued to it, to form a completed, open hull ready for fitting out.

Finally, the hull can be built as per the original 1:1 version. In that case, a balsa plug the shape of the hull but undersize by the thickness of the hull, is first made and then the framework of ribs, keel and topwales built over it (care being taken not to let them stick to it). Once finished, this skeleton should be capable, with a little prising, of being lifted off. The surface of the plug is reduced to clear the gaps in it and the skeleton replaced when it is finally planked with strips of lime, just like a real ship.

Bases and cases

A display case, cupboard, box or similar really is a must; the modeller's greatest enemy is dust. Models, especially those with rigging and tiny, delicate details, attract it like the proverbial fluff to a lollipop, but are almost impossible to keep dusted without causing damage. The more airtight the case, the better. For full-hull models, a simple stand is desirable; this need be no more than a couple of hull section negatives joined by a short bar. For the very small, light models, the author uses clear cellophane (as found in the packaging for shirts) for this as it produces a virtually invisible stand that does not detract from the model. Waterline models can be set on bases, either plain, usually best finished in a complementary colour, or scenic. In the latter case, 'water' can be fashioned using fine surface filler, worked with a palette knife or similar. Do bear in mind that if the model is simply put on the base and the 'sea' worked around it, the ship will appear too low and the model should thus first be mounted on a piece

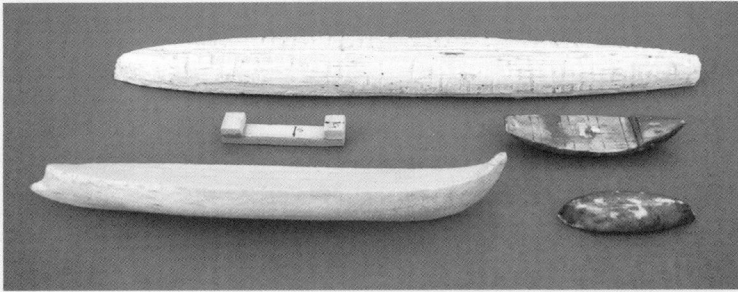

Left, moulds. The upper mould was used for a model lusoria at 1:96 scale and the marks can be seen where ribs and stringers were placed, prior to planking the hull. Unlike the 'plug' method, here the planking forms the structural skin of the model and is used for open, undecked boats. The straight piece below it is one of the retainers, made to fit over the top wales to hold them on to the mould while the hull is planked. Two moulds for small, open boats are below with another, as yet unused mould for a scout ship.

Right, plugs. For this full-hull model of a quinquereme at 1:192 scale, layers of balsa sheet have been screwed together and the hull roughly shaped. The top layer will be removed once the hull shaping is finished, for recesses to be made to form the deck voids below hatches and for side galleries behind ventilation screens. The layers can then be permanently glued to form the basic hull, which will then be 'planked' with strips of lime.

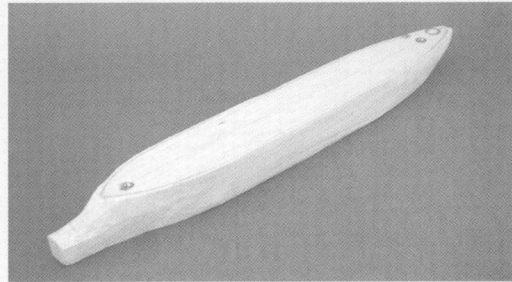

of wood the thickness of the proposed seascape. Watercolours can be used for the sea to give a realistic, dull finish; they enable the complex mix of blues and greens to be worked in to reproduce the elusive 'colour of water'. Bow and stern waves can be highlighted afterwards with a little white matt enamel.

On the matter of scenic settings, another advantage of 1:300th scale is that there is a splendid range of wargames figures available in that scale which can be used as crew. Of course, there are other scale figures also available, such as 1/72nd, and various wargaming and railway scales come to mind, but their use will demand bigger models to suit

Some other details

For small-scale modelling as envisaged here, shavings of lime have a multitude of uses. Using a jack plane with, of course, a keen blade, shavings can be taken from your piece of limewood; they will come out silky smooth on one side and barely rough on the other. Carefully unrolled, they can be stored between two pieces of flat wood held by rubber bands (like a sandwich) until needed. Kept like this the shavings will gradually lose much of their curl and be easier to use. With a little practice, shavings can be taken in varying thicknesses, even thin enough for light to shine through.

The shavings can be cut into strips for deck and hull planking, bulwarks, cabin sides and roofs, seats, rails; they can be laminated, bent and twisted and always need only the lightest of finishing to give a fine, smooth surface, either for painting or to be left in its natural state.

To make open hatches in block or solid hulls, the appropriate wood should be cut away and the voids hollowed out and squared so that they are slightly bigger than the hatches and then painted matt black. A piece of lime shaving is cut to the finished deck plan and the hatch openings accurately cut out. The deck is attached to the hull and tiny strips fitted around the hatch edges to form coamings. The same basic process is also used for open courses or galleries in the sides of hulls. Making such holes oversized behind the covering edges of their finished openings, gives the impression of an open interior of the hull.

For making wales semi-circular in section for hull sides, plastic insulation of suitable diameter can be stripped from pieces of electrical wire and cut in half lengthways. Slit along one side, it can be fitted over the edge of bulwarks to form cappings.

The oars of the small models are formed by cutting suitable lengths of 15-amp fuse wire or telephone wire and flattening one end by squeezing with pliers; the ends are then squared to produce an oar blade. To straighten the wire ready for such use, fix one end of the length in a vice and, grabbing the other end with pliers, pull and stretch the wire, or use two pairs of pliers if you are feeling particularly muscular.

If making buildings, quays and the like for a diorama setting for a model, blocks of wood smeared with a layer of fine surface filler, which gives the right texture and can be scored to resemble brick or stone work, can be used. The filler will take watercolour well for a realistic finish and ordinary pencil can be lightly used for marking details.

These methods and techniques are not intended to be exhaustive. With practice and experience every modeller evolves their own favourite methods and usages; there are always new things to be tried, along with new challenges and problems to be overcome, but then surely that is part of what makes it so interesting.

There are many books on modelling boats and ships but comparatively few that concentrate on miniatures of a nature comparable to Roman warships. A selection of useful titles is: Freeston, *Model Open Boats*, Conway 1975; McNarry, *Shipbuilding in Miniature*, Conway 1982; McCaffery, *Ships in Miniature*, Conway 1988; and of course, Pitassi, *Roman Warships*, Boydell 2011.

Examples (see colour plate section)

Half-model, monoreme penteconter, scale 1:192

The hull of this model was from a block of balsa wood, cut to the overall length, beam and depth of the hull (that is the bottom of the keel to the highest point of the upper deck) and squared. The profile of the bulwark sheer (excluding bow and stern rails) was marked and cut. Next a centre line was marked all around the block and the whole deck plan, to its widest extent marked on the top (deck) and bottom of the block and cut. The result was now a block, flat on the bottom, with the upper deck sheer cut on the top and slab-sided hull, formed to its final plan. With the basic parameters set, the hull was next be carved and shaped to its final, finished form and given a coat of knotting to tighten-up and harden the surface. Carefully cutting along the centre line all around to separate the halves, each of which was next hollowed with knife and gouges to form a shell about 1/16 inch (1.5 mm) thick. The original intention had been to fit both halves to a keel to make a whole hull but a slip here and there meant that one half was not good enough and discarded. The half model was the result but did offer the compensation of the opportunity to put extra detail within it which could be displayed. The keel, stem and stern profiles were fashioned from 1/32nd inch ply but any suitable thin timber will do. The hull half was given a coat of knotting inside and lightly rubbed down, to give an even colour finish. Ribs, wales, struts and other timbers were added from strips of lighter-coloured wood to give contrast; rigging from copper wire (does not sag). All the other paraphernalia, such as water jars, spears, shields, spare oars, rope coils and such like were inserted, together with a crew, made from twisted wire armatures, covered in plasticene and painted. The whole, completed model was then stuck to a sheet of glass, inscribed with the waterline and boxed in.

A late Roman warship, scale 1:300

For this small, full-hull model, once again balsa wood was used for the hull. The whole block was marked and carved as before but then a shallow slot was cut for the keel, stem and stern posts and slots cut out along the sides for the rower's gallery and sections of the top deck cut away below where the hatches would be, before it was knotted and rubbed down. The interior cut-away voids were painted matt black. The

Modelling Techniques

keel, stem and stern posts were made from lime
and fitted and a complete upper deck made
from a piece of lime shaving bigger all around
than the hull block, to allow for final trimming
once stuck down; the hatch openings were
carefully cut out and the mast position marked
before fixing. As they were to be painted, the
side panels were cut from fine card and painted
inside and out with watercolours before being
attached in place. Stanchions made from slivers
of lime could then be fitted inside the bulwarks
and wales, also from lime, to the outside.
Rudders and tiller were also from lime and their
mounting brackets from pieces of wine-bottle
foil. The mast and spars are from shaved-down
dowel, the sail from paper, suitably tinted with
watercolour and slightly crumpled around the
yard to represent furling. The rigging is made
from fine wire and secured to suitably placed
deck rings, bent from slightly thicker brass
wire. Finally a couple of anchors were made
from wood bound by strips of foil and fixed,
with coils of wire 'rope' to the foredeck. Oars
from soft (15 amp) wire flattened at the ends,
completed the model.

River liburnian, scale 1:96

This was made from a block of pear wood with
the centre line marked. The keel, stem and stern
posts were made up from oak, with a section
of the keel extended upward, into what would
become the hull void. The hull block was cut
through on the centre line and the keel assembly
fixed with short dowels through its extended
section and a short way into the hull halves, to
join them into a single hull. The hull was then
carved to its final shape, the halves being able to
be removed for matching. Once this was done,
each half-hull could be hollowed out in turn to
form the shell and when completed, affixed to
the keel assembly to make the completed hull.
Everything else then is built up and attaches to
the hull.

Celox, scale 1:96

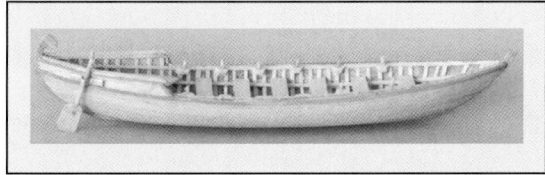

This model was made using an amalgam of
the other methods. The whole hull, including
the raised stern part was carved from a block
of balsa wood, slightly undersize, knotted and
rubbed down. Next, keel stem and stern pieces
were fashioned from pear wood and glued into
appropriate slots cut into the hull block, with
PVA.

The outside of the hull was then completely
planked with strips of lime shaving, again using
PVA, trimmed and rubbed down to its final
form. A coating of PVA was applied to the
whole outside of the hull to strengthen it before
the inside was hollowed out to form a thin shell
of balsa and planking. The finished hull shell is
only about 1/32nd inch thick (1 mm.), although
the bottom was left a little thicker as it was to
be covered by bottom boards and also had to
clear the depth of the keel insert. After knotting
and smoothing the inside, interior detail, ribs,
thwarts, carlings, tholes and bottom boards
were added, made from strips of lime. The
exterior decorative strip was from thin card,
washed with watercolour and the side rudder
bracket from metal foil.

With all of these ships, the hull is the most
complex thing and extra attention paid to making
it as authentic as possible and especially so that
it 'looks right,' that is that a full-size version,
while conforming to the available iconography,
would sit in the water well and actually work in
practice, while giving that essential impression
of *Romanitas*. It can be seen that everything
else, once the hull is right, becomes a matter of
embellishment, small detail added to give depth
to the model and to whatever degree is desired
by the modeller. In the absence of surviving
detailed descriptions, this area leaves scope for
interpretation, subject always of course, to the
materials and technology of its time and limited
to what would have been necessary or desirable.

THE MODEL IS TO BE of the later type of quinquereme, with finer, more refined lines. The sequence of manufacture is shown in the numbered photographs.

1 The hull shape has been carved from a single block of balsa, cut from a sheet of appropriate thickness, upper deck sheer at bow and stern cut and the level centre part reduced to its correct scale height above the waterline, known to be 10 Roman feet (9 feet 3 inches or 296 mm). Recesses have been hollowed out for the voids beneath the deck hatch openings. More shaping to finalise the bow and ram is needed.

2 The hull has had a coat of knotting, which has been rubbed down. Slots have been cut into the centre line at bow and stern to accept the stem and stern posts, which have been shaped from slices of lime, shown lying below the hull block. Between them can be seen the pieces, made from basswood, that will form the port and starboard oarboxes; they have been bevelled on the top, save for the aftermost section, which will house the steering oars and the position of the thranite oar ports marked with a pin. Finally they have been matched to the hull shape of their respective sides and are ready for fixing. Recesses have been cut from each side, level with the tops of the oarboxes, which will be covered by the ventilation course. Below the oarbox position, narrow slots have been cut for the openings of the zygite reme. Also seen are a couple of strips of lime shaving that will be used for planking the hull block.

3 The recesses for the thranite and zygite remes, together with the deck hatch recesses, have been painted matt black. Stem and sternposts and the oarboxes have been fixed using PVA glue and the ram finally shaped. A thicker lime shaving has been prepared for the deck covering (laying above the hull) with notches to fit around the stem and stern posts and the deck hatches cut out; final shaping to match the hull form will be done once it is

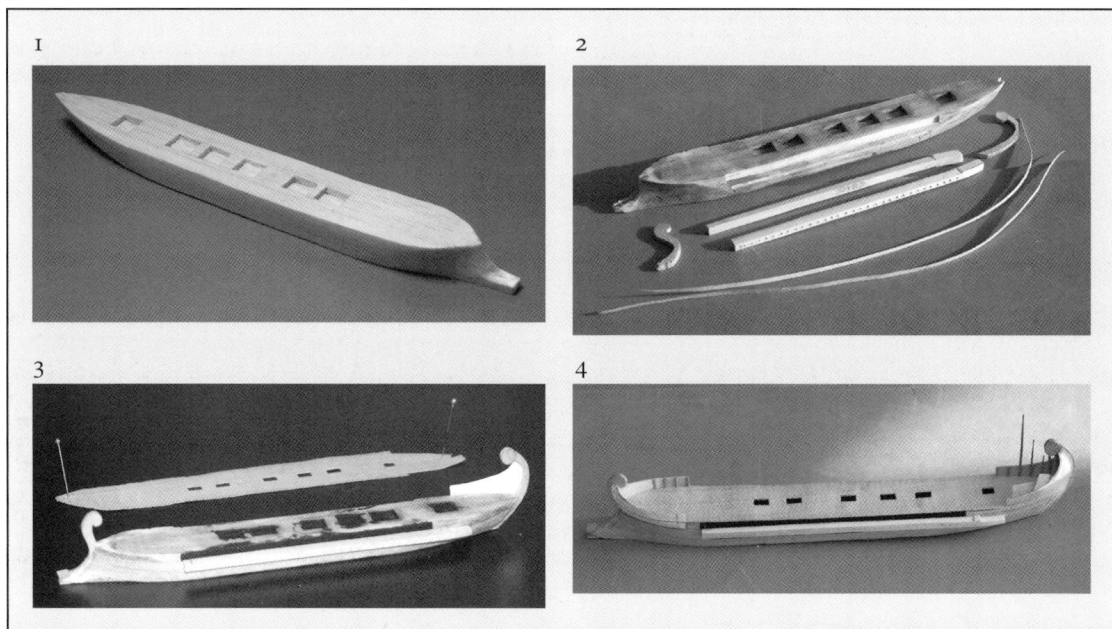

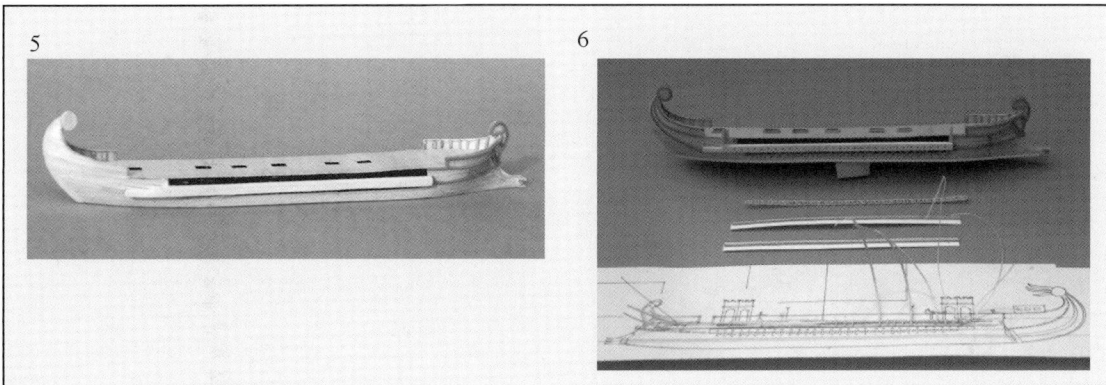

affixed, which will be done once the hull body planking is completed. Planking of the hull has commenced (note the way the planking is swept up at bow and stern) using PVA glue. For illustration, one of the paper stern bulwark templates has been fitted; normally this would be done after the deck has been fitted and trimmed to shape.

4 Hull planking has been completed, the deck has been secured and trimmed to match the hull sides. The paper templates forming the bow and stern bulwarks have been fitted and are in the process of being planked inside and out. The portside stern bulwark is partially planked outside, the strakes running in alignment with the hull planking, as if a continuation of it; inside, because of the concave curves, short lengths of shaving are attached vertically and trimmed once set. Tiny strips in a contrasting wood (here, pear) are fashioned and fixed to form stanchions inside the bulwarks, as can be seen on the starboard side.

5 The hull and bulwark planking is completed, the ram finished and decoration is being added. To make these strips, lengths of insulation from fine wire 1/32 inch (1 mm) diameter is split lengthways and glued into position, using cyanoacrilate and approximately following a pattern shown on coins. This material has the advantages of being uniformly pre-coloured and completely flexible and able to conform to the curves of the ship, including tight turns as on the stemhead.

6 The hull decoration is completed and a waterline armour belt of thin, plastic card,

added. That this is matt white could represent white paint or a coating of lime and wax, which the Romans used to deter marine parasites. The oarports for the lowest, thalamite oars have been marked with a pin and stanchions have been added to support the outrigger and to divide and define the zygite reme. Coamings of thin, flat strips of lime have been fixed around the deck hatches and are being trimmed. Deck edge bulwarks from thin card and wood strip are shown laid above the plan and will be fitted after the ventilation course grilles. One of these can be seen between the bulwarks and the hull; the other is being made from very thin strips of lime, using a copy of the elevation as a guide. It is a very fiddly process at these small scales but when finished, adds a whole dimension of detail to the model.

7 Bulwarks (suitably tinted with watercolour before fitting) and the ventilation grilles have been fitted and stanchions put inside the bulwarks to complete them. A cap rail from lime strip has been added to them and also to the bow and stern bulwarks. While access is still

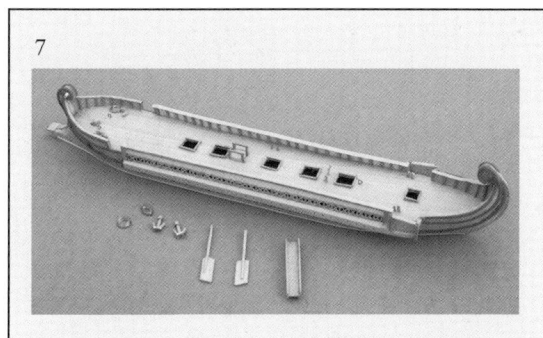

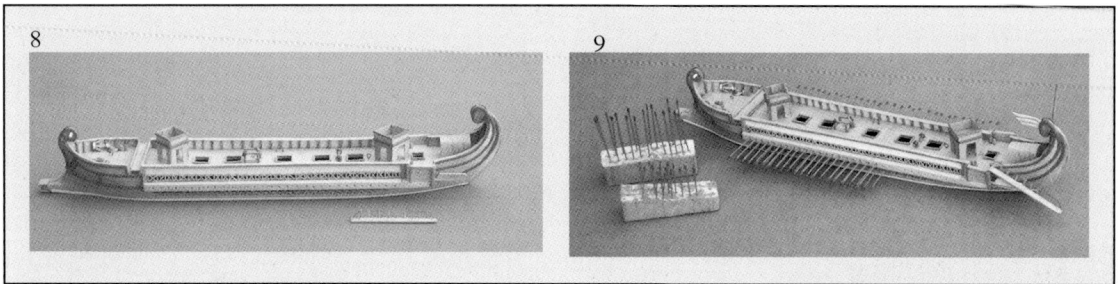

easy, small deck fittings have been added: bitts, a deck shoe and bracket on the stempost for an *artemon* and a mast tabernacle together with deck rings for the rigging made from twisted brass wire. The framework and panelling to form the side-rudder housings has been completed. In the foreground, are the rudders, made from pieces of lime and a boarding bridge, also ready for fitting aboard. Anchors have been made from pieces of lime bound with strips of metal foil and coils of rope from copper wire will be placed (secured with glue) around the mast and anchors.

8 Rudder brackets from metal foil have been added to the rear of the rudder sponsons but, as the ship is intended to be shown hove-to the rudders have been mounted in the raised position, as shown in several ancient reliefs. Fighting towers have been made from two sheets of lime shaving, laminated with the grain of one layer at ninety degrees to the other, to avoid warping. The four sides of each tower is cut out and then assembled around a square piece, representing the floor and also stuck along the corners.

They can also be made from paper, if the intention is to paint them, which is a little less fiddly, but here the natural wood finish was desired. The stern cabin has been made from paper, coloured to resemble canvas, over wood formers. The gangplank (in front of the model) is from pieces of lime with a fine wire hand rope

9 The gangplank is shown slung from the port-side rudder sponson (as shown in reliefs) and a small spar in the upper stern, it is shown lowered. A signal staff made from a fine strip of lime and sanded to a slight taper, has been added in the stern and decorative vanes from painted plastic card complete the stern-post decoration. The ram is completed by being faired into the hull and stem, with a sheathing of metal foil and painted bronze. The 176 oars are made from 15-amp or telephone wire, flattened at the end and painted. They are pushed into place in their respective remes where previously marked with a pin, they are in the at ease position, to complete the ship model.

10 As the model is intended for a diorama, four artillery pieces and a deck crew (eight officers, twenty sailors and forty marines) are painted and mounted on the deck. For this particular scene, an eight-oar celox has been carved from balsa and fitted with pieces of lime shaving, again with a suitable crew (eight

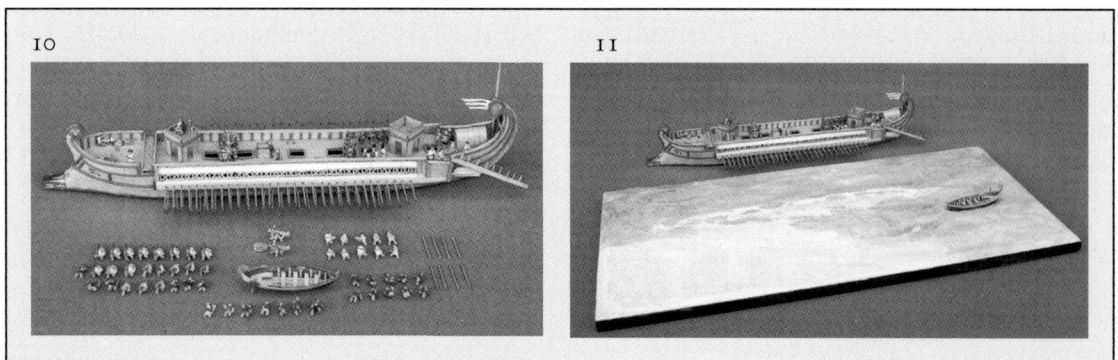

12

rowers, bow and helmsman) and figures painted for the prefect and his adjutant (eighty figures in all, all figures by Heroics and Ros Figures of Trowbridge). Some of the figures have to be adapted to become the rowers in the celox and sailors in plain tunics.

11 The base for the diorama (here 9 by 4½ inches (228 by 114 mm.) is from a non-warping material, here plywood, but a fibre or chipboard would also suffice. The models are positioned and elevated, to allow for the thickness of the sea so that they will float to their correct waterlines when it is added. The models can then be removed so the sea can be worked on. This is formed from fine surface filler, suitably

fashioned and painted with watercolours, for a flat finish. The edges of the base are painted matt black.

12 With the base completed, the models are secured in their positions. These ships are hove to but for one under way, bow waves, wakes and oar splashes can be added with a little extra white filler after mounting. Finally, the modeller's greatest enemy, dust, must be opposed and a suitable cover provided. The diorama has been secured to another base with a moulded edge, which has been varnished. A cover has been made from acrylic sheet, to fit into the slot between the mounting and the moulding and to cover the whole.

REFERENCES AND NOTES

PRIMARY SOURCES

Ammianus Marcellinus, *Roman History*. Trans: C. D. Yonge
Apicius, *The Roman Cookery of Apicius*. Trans: J. Edwards
Appian, *The Civil Wars*. Trans: John Carter
Appian, *The Illyrian Wars*. Trans: H. White
Appian, *The Mithridatic Wars*. Trans: H. White
Athenaeus. Trans: C. D. Yonge
Aurelius Victor, *De Caesaribus*. Trans: H. W. Bird
Caesar, *The Battle for Gaul*. Trans: A. and P. Wiseman
Caesar, *The Civil War*. Trans: F. P. Long
Cicero, *Ad Atticus, Letters to Atticus*. Trans: E. S. Shuckburgh
Codex Theodosianus. Trans/ed: T. Mommsen
Dio Cassius, *Roman History*. Trans: I. Scott-Kilvert and J. Carter
Diodorus Siculus. Trans: G. Booth
Herodotus, *The Histories*. Trans: G. Rawlinson
Hirtius, *The Alexandrine War*. Quoted in Torr, *Ancient Ships*
Homer, *The Iliad*. Trans: E. V. Rieu
Lives of the Later Caesars, First Part of the Augustan History, Trans: A. Birley
Livy, *The History of Rome*. Trans: A. De Selincourt, D. Spillan and C. Edmonds
Lucan, *The Civil War*. Trans: D. B. Killings
Orosius, *History*. Quoted in Torr, *Ancient Ships*
Pliny, *Natural History*. Trans: J. Healey
Plutarch, *Lives: Pompeius*. Trans: G. Long
Plutarch, *Lives: Marcus Cato*. Trans: G. Long
Procopius, *The Vandalic wars*. Trans: H. B. Dewing
Procopius, *The Persian Wars*. Trans: H. B. Dewing
Polybius, *The Rise of the Roman Empire*. Trans: I. Scott-Kilvert
Silius Italicus, *Punica*. Trans: J. D. Duff
Suetonius, *The Twelve Caesars; Augustus*. Trans: A. Thomson
Tacitus, *Agricola* and *Germania*. Trans: H. Mattingly
Tacitus, *The Annals*. Trans: M. Grant
Tacitus, *The Histories*. Trans: K. Wellesley
Theophrastus, *Historia Plantarum*. Quoted in Welsh, *Building the Trireme*
Thucydides, *The Peloponnesian Wars*. Trans: B. Jowett
Vegetius, *Epitome of Military Science*. Trans: N. P. Milner
Xenophon, *The Persian Expedition*. Trans: R. Warner
Zosimus, *New History*. Trans: R. T. Ridley

OTHER SOURCES

Asskamp and Schafer *Projekt Romerschiff* Koehlers Verlagsgesellschaft 2008
Austin and Rankov *Exploratio, Military and Political Intelligence in the Roman World* Routledge 1995
Barker and Rasmussen *The Etruscans* Blackwell Publishers 1998
Baines and Malek *Atlas of Ancient Egypt* Phaidon Press 1980
Bass G.F. *A History of Seafaring Based on Underwater Archaeology* Thames & Hudon 1972
Bennett G. *The Battle of Jutland* Batsford 1964
Berthold R. M. *Rhodes in the Hellenistic Age* Cornell University Press 1984
Bickerman E.J. *Chronology of the Ancient World* Thames & Hudson 1980
Bishop and Coulston *Roman Military Equipment* Second edition, Oxbow Books 2006
Bounegru and Zahariade *Les Forces Navales du Bas Danube et de la Mer Noir* Oxbow Books 1996
Bradford E. *Ulysses Found* Hodder & Stoughton 1963
Burn A.R. *The Pelican History of Greece* Penguin Books 1966
Casson L. *Libraries in the Ancient World* Yale University Press 2002
Casson L. *Ships and Seamanship in the Ancient World* Princeton University Press 1986
Casson L. *The Ancient Mariners* Second edition, Princeton University Press 1991
Casson L. *Travel in the Ancient World* Johns Hopkins University Press 1994
Casson and Steffy *The Athlit Ram* Texas A & M University Press 1991
Champion J. *The Tyrants of Syracuse* Pen & Sword Books 2010
Christensen A. E. *Proto-Viking, Viking and Norse Craft* in *The Earliest Ships*, Conway Maritime Press 2004
Christensen A. E. *Scandinavian Ships from Earliest Times to the Vikings* in Bass, *A History of Seafaring Based on Underwater Archaeology*
Cowell F. R. *Everyday Life in Ancient Rome* Batsford 1961
Cunliffe B. *The Extraordinary Voyage of Pytheas the Greek* Penguin Books 2002
Duncan D. E. *The Calendar* Fourth Estate 1998
Ellmers D. *Celtic Plank Boats and Ships, 500 BC–AD1000* in *The Earliest Ships*, Conway Maritime Press 2004
Elton H. *Warfare in Roman Europe, AD 350–425* Oxford University Press 1997
Feugere M. *Weapons of the Romans* Tempus 2002
Frost H. 'The Punic Wreck in Sicily' *International Journal of Naval Archaeology* 1974
Green P. *Alexander to Actium* University of California Press 1990
Guerber H. A. *The Myths of Greece and Rome* Harrap 1938
Guilmartin J. E. *Gunpowder and Galleys* Conway 2003
Harden D. *The Phoenicians* Thames & Hudson 1962

Haywood J. *Dark Age Naval Power* Anglo-Saxon Books 1999

Heather P. *Empires and Barbarians* Pan Books 2010

James and Thorpe *Ancient Inventions* Ballantine Books 1994

Johansen R. *The Viking Ships of Skuldelev* in *Sailing into the Past.* Seaforth Publishing 2009

Jones A. H. M. *Augustus* Chatto & Windus 1970

Larousse *Encyclopaedia of Mythology* Paul Hamlyn 1964

Luttwak E. N. *The Grand Strategy of the Roman Empire* Johns Hopkins University Press 1976

Marlow J. *The Golden Age of Alexandria* Victor Gollancz 1971

Marsden E. W. *Greek and Roman Artillery – Historical Development* Oxford University Press 1969

Marsden E. W. *Greek and Roman Artillery – Technical Treatises* Oxford University Press 1969

McGrail S. *Ancient Boats and Ships* Shire Archaeology 2006

Morrison and Coates *Greek and Roman Oared Warships, 399–30 BC* Oxbow Books 1996

Morrison, Coates and Rankov *The Athenian Trireme* Second edition, Cambridge University Press 2000

Moscati S. *The World of the Phoenicians* Sphere Books 1973

Newark T. *The Barbarian, Warriors and Wars of the Dark Ages* Blandford Press 1985

Norwich J. J. *A History of Venice* Penguin Books 1982

Norwich J. J. *Byzantium. The Early Centuries* Penguin 1988

Ormerod H. A. *Piracy in the Ancient World* Dorset Press 1987

Patai R. *The Children of Noah. Jewish Seafaring in Ancient Time*s Princeton University Press 1999

Pearson A. *The Roman Shore Forts* Tempus Publishing 2002

Perowne S. *The Life and Times of Herod the Great* Sutton Publishing 2003

Pitassi M. *Roman Warships* The Boydell Press 2011

Pitassi M. *The Navies of Rome* The Boydell Press 2009

Pryor J. H. *The Geographical Conditions of Galley Navigation in the Mediterranean.* in *The Age of the Galley.* Conway Maritime Press 1995

Rankov B. *The Greek Trireme Olympias* in *Sailing into the Past.* Seaforth Publishing 2009

Rodgers W. L. *Greek and Roman Naval Warfare* U.S. Naval Institute 1937

Rogan J. *Reading Roman Inscriptions* Tempus 2006

Rogerson J. *Chronicle of the Old Testament Kings* Thames & Hudson 1999

Salmon E. T. *Samnium and the Samnites* Cambridge University Press 1967

Scullard H. H. *A History of the Roman World, 753–146 BC* Fourth edition, Routledge 1980

Scullard H. H. *From the Gracchi to Nero, a History of Rome, 133 BC to AD 68* Fifth edition, Routledge 1982

Selkirk R. *The Piercebridge Formula* Patrick Stephens 1983

Severin T. *The Jason Voyage* Hutchinson 1985

Severin T. *The Ulysses Voyage* Hutchinson 1987

Shepard A. M. *Sea Power in Ancient History* William Heinemann 1925

Shepherd S. *Pickled, Potted and Canned, how the preservation of food changed civilisation.* Headline Book Publishing 2001

Sitwell N. H. H. *The World the Romans Knew* Hamish Hamilton 1984

Sobel D. *Longitude* Fourth Estate 1996

Sprague de Camp L. *The Ancient Engineers* Ballantine 1963

Starr C. G. *The Roman Imperial Navy, 31 BC to AD 324* Third edition, Ares Publishers 1993

Starr C. G. *The Influence of Seapower on Ancient History* Oxford University Press 1989

Tannahill R. *Food in History* Stein & Day 1973

Torr C. *Ancient Ships* Argonaut 1964

Warry J. *Warfare in the Classical World* Salamander Books 1980

Websites:

　Navis Online Database of Ancient Ships

　Oxford Roman Economy Project, Online database of Shipwrecks

Welsh F. *Building the Trireme* Constable 1988

Wilkins A. *Roman Artillery* Shire Archaeology 2003

NOTES

I   A BRIEF HISTORY OF THE ROMAN NAVY

1　Livy, V.34
2　Polybius, III.22; Livy, IX.43
3　Livy, V.28
4　Livy, VII.25/26
5　Livy, VIII.14
6　Dio Cassius, IX, frag. 39, mentions a trireme, *Roman History*
7　Livy, IX.28
8　Scullard, *A History of the Roman World*
9　Livy, IX.38
10　Polybius, III.25
11　Dio Cassius, *Roman History*, IX (frag. XXXIX)
12　Salmon, *Samnium and the Samnites*
13　Harden, *The Phoenicians*; Moscati, *The World of the Phoenicians*
14　Bradford, *Ulysses Found*, for personal experience of sailing in these waters
15　Polybius, I. 20
16　Polybius, I.22
17　Livy XXI.50
18　Polybius, I.39, who exaggerates the losses
19　Polybius, I.61
20　Polybius, I.63
21　Polybius, III.41; Livy, XXI.17
22　Livy, XXIV.11
23　Livy, XXXVI.45
24　Livy, XXVIII
25　Livy, XXVII.23/24
26　Ormerod, *Piracy in the Ancient World*
27　Plutarch, *Pompeius* XXIV–XXVIII
28　Caesar, *The Battle for Gaul*, III.14/15
29　Caesar, *The Battle for Gaul*, IV.22/25

30 Caesar, *Civil Wars* and Appian, *The Civil Wars*, are the principal sources.

31 Caesar, *Civil War* III

32 Berthold, *Rhodes in the Hellenistic Age*

33 Tacitus, *The Annals* XI.14–19

34 Tacitus, *Histories* IV.79

35 Tacitus, *Histories* V.22

36 The revolt of Civilis and the campaign against him are covered in books IV and V of Tacitus' *Histories*

37 Aurelius Victor, *De Caesaribus* 39

38 Luttwak, *The Grand Strategy*

39 Ammianus Marcellinus XVII.2

40 Zosimus, IV.38/39

41 Aurelius Victor, *De Caesaribus* 33

42 Procopius, *The Vandal Wars* III

43 Procopius, *The Vandal Wars* III.6/22/27

2 SHIPBUILDING

1 There are databases of ancient ship finds online: see under websites in the list of other sources

2 Of the several ships discovered at the site of what was the ancient port of Pisa, Ship 'C' has a ram-type prow and six surviving rowing benches, together with oarports with integral metal tholes. Date still uncertain but it is (only) a possibility that this is a small warship

3 The originals and replicas are housed at the Museum of Ancient Shipping at Mainz

4 The Cape Gelidonya wreck; Bass, *A History of Seafaring*

5 Bass, *A History of Seafaring*; Casson, *Ships and Seamanship*

6 Modern examples of this method of construction include the Argo: Severin, *The Jason Voyage* and the trireme Olympias; Morrison, Coates and Rankov, *The Athenian Trireme*; Welsh, *Building the Trireme*

7 Casson, *Ships and Seamanship*

8 Casson, *Ships and Seamanship*; Bass *A History of Seafaring*. The Yassi Ada wreck (fourth century AD) distinguished from the Pantano Longarini ship (post-fifth century AD)

9 Vegetius, *Epitome of Military Science* IV.34, recommends bronze rather than iron fixings

10 Pitassi, *Roman Warships*

11 Procopius, *The Persian Wars* I.19; Vegetius, *Epitome of Military Science* IV.37

12 Livy XXIX.1

13 Casson and Steffy, *The Athlit Ram*

14 H. Frost, 'The Punic Wrecks in Sicily'

15 Caesar, *The Battle for Gaul* II.13

16 Ellmers, *Celtic Plank Boats and Ships*, in Conway's *The Earliest Ships*; McGrail, *Ancient Boats and Ships*

17 The mid-second-century 'Blackfriars ship' was built frame first whereas the late third century 'County Hall ship' was built in southern England, using the mortise and tenon method; both found in London

18 Theophrastus V.7.2

19 Bass, *A History of Seafaring*

20 Herodotus II.96

21 Livy XXIX.1

22 Caesar, *Civil War* III.58

23 Vegetius, *Epitome of Military Science* IV.36

24 The *Argo*, Severin, *The Jason Voyage*; the replica trireme *Olympias*

25 Theophrastus IV.1.2; Pliny, *Natural History* XVI.195

26 Theophrastus V.1.6; Homer, *The Iliad* VII.5

27 Livy XXVI.47

28 Vegetius, *Epitome of Military Science* IV.37

29 Morrison and Coates, *Greek and Roman Oared Warships*

30 Morrison and Coates, *Greek and Roman Oared Warships*

31 The two ships found off Marsala have shipwright's marks cut into the timbers to aid assembly, rather than their having been built together

32 Livy XXVIII.45

33 Polybius I.20/21

34 Livy XXVIII.45

35 Pitassi, *The Navies of Rome*

36 Vegetius, *Epitome of Military Science* IV.46

37 *Codex Theodosianus* VII.17

3 SHIPBOARD WEAPONS

1 For example, Caesar, *Civil War* II.1.6,7

2 Morrison, Coates and Rankov, *The Athenian Trireme*

3 Livy XXII.20

4 These ships, fount near Motya, were captured and being used as blockships by the Romans, trying (unsuccessfully) to block the harbour of Marsala which they were besieging.

5 Probably the best example being a complete ram found off the Israeli coast and thought to have been from a quadrireme; see Casson and Steffy, *The Athlit Ram*

6 Morrison and Coates, *Greek and Roman Oared warships*

7 The type can be seen on warships shown on Trajan's Column in Rome

8 Pitassi, *Roman Warships*

9 Polybius XVI.3

10 Vegetius, *Epitome of Military Science* IV.37

11 Livy XXXVII.11; Polybius XXI.7

12 Appian, *De Bello Civilibus* IV.115/116, quoted in Shepard, *Sea Power in Ancient History*

13 Cassius Dio L.43/35

14 Vegetius, *Epitome of Military Science* IV.44

15 Thucydides, *The Peloponnesian Wars* VII.53

16 Caesar, *The Civil War* III.5.101

17 Procopius, *The Vandal Wars* III.6.22-27

18 Thucydides, *The Peloponnesian Wars* VII.25

19 Polybius VIII.4

20 Pitassi, *Roman Warships*

21 Polybius XVI.2-15; this full account of this battle is unique and probably the best description of a battle between Hellenistic polyremes and the confusion involved, as well as the intensity of the fighting

22 Pliny, *Natural History* XXXII

23 Polybius I.22

24 Polybius I.22

25 Polybius I.61; Vegetius, *Epitome of Military Science* IV.44

26 Marsden, *Greek and Roman Artillery*

27 The development of the high-explosive shell provided

the first such weapon

28 As at Actium and the final action of the 'Civilis revolt' in AD 69, Tacitus, *The Histories* V.23

29 Caesar, *The Battle for Gaul* IV.25

30 Caesar, *The Alexandrine War* XIX.3

31 Tacitus, *The Annals* XV.7

32 Marsden, *Greek and Roman Artillery*

33 Livy XXVI.26

34 Livy XXVII.15

35 Polybius X.12

36 Livy XXX.4

37 For example, Caesar, *Civil War* II.1.6

38 Appian, *Civil Wars* V.118/119

39 Appian, *Civil Wars* VII.8

40 Athenaeus *Deipnosophistae* V.43, describes a great ship built at Syracuse in about 265 BC as having 'eight towers of the same height as the deck structures of the ship', i.e. platforms

41 Marsden, *Greek and Roman Artillery*; Wilkins, *Roman Artillery*

4 TYPES OF SHIP – THE REPUBLIC

1 After Assyrian reliefs from Ninevah (now in the British Museum) of that time

2 Thucydides, *The Peloponnesian Wars* I.13

3 For one among many examples, see Herodotus, *The Histories* I.52

4 Casson, *Ships and Seamanship*

5 Livy I.23

6 Polybius I.20

7 Diodorus Siculus XIV.42

8 Polybius I.20

9 Pitassi, *Roman Warships*

10 Orosius, VI.19

11 Polybius I.26

12 Polybius XV.2

13 Pliny VII.207

14 As discussed, Pitassi, *Roman Warships*

15 Polybius I.26; Livy XXIX

16 Polybius I.23

17 Appian, *The Illyrian Wars* III

18 Livy XXIV.35

19 Lucan, *The Civil War* III.534

5 TYPES OF SHIP – THE EMPIRE

1 The types are well attested on grave stelae; Morrison and Coates, *Greek and Roman Oared Warships*

2 Zosimus I.71

3 For a discussion of this battle and the incongruities of the surviving accounts, Pitassi, *The Navies of Rome*

4 Pitassi, *Roman Warships*

5 Pitassi, *Roman Warships*

6 The date of the remains of two found at Oberstimm near Ingolstadt in Bavaria, now in the Museum of Ancient Shipping, Mainz

7 Asskamp and Schafer, *Projekt Romerschiff*, for a full account of the construction and sailing of a replica twenty-oared 'Oberstimm' boat

8 Caesar, *The Battle for Gaul* III.13

9 Vegetius, *Epitome of Military Science* IV.37

10 These boats, together with reconstructions of them are at the Museum of Ancient Shipping, Mainz

11 Vegetius, *Epitome of Military Science* II.1, gives lusoriae as 'river patrol boats' but distinguishes them from 'warships' which he calls liburnae

12 Codex Theodosianus VII.17

13 Pitassi, *Roman Warships*

14 Procopius, *The Vandalic Wars* I.11

6 ORGANISATION OF THE FLEETS

1 For a more detailed view of the political arrangements, see Scullard H. H., *A History of the Roman World*

2 Ibid.

3 Polybius I.49

4 Polybius I.20

5 Polybius I.25, although modern estimates put the number at about 230 ships

6 Polybius I.41; Livy XXI.17

7 Caesar, *Civil War* III.1.3; Shepard A. M., *Sea Power in Ancient History*

8 Tacitus, *The Annals* IV.3

9 C. G. Starr, *The Roman Imperial Navy*

10 Ibid.

11 Tacitus, *The Histories* I.6

12 As the spheres of operation, see Vegetius, *Epitome of Military Science* IV.31, although he is commenting on a situation which was already in the past

13 Starr, *The Roman Imperial Navy*

14 Nonius Datus, quoted in Sprague de Camp, *The Ancient Engineers*

15 Starr, *The Roman Imperial Navy*; the poet Horace journeyed along this coast in the first century BC

16 Tacitus, *The Annals*, I.8 et seq.

17 It is known from only one inscription (so far), namely a grave stele to a member of this 'fleet' from the reign of Domitian (AD 81–96), perhaps therefore a veteran of this campaign. Starr, *The Roman Imperial Navy*

18 Starr, *The Roman Imperial Navy*

19 As an example, at random, in 341 BC, the Senate provided soldiers of the army with a year's pay and corn for three months, upon withdrawal from active service. Livy, VIII.2

20 Polybius, I.59

21 Livy, XXIV.11

22 Livy, XXXVI.35

23 Livy, XXVIII.45

24 Pitassi, *The Navies of Rome*

25 Elton, *Warfare in Roman Europe, AD 350–425*. Procopius, *The Vandal Wars*, III.6

7 THE COMMAND STRUCTURE

1 Livy IX.38; X.2

2 Livy XXI.17

3 Livy XXIV.40

4 Ormerod, *Piracy in the Ancient World*

5 Scullard, *A History of the Roman World*

6 Starr, *The Roman Imperial Navy*

7 Procopius, *The Vandal Wars* III.6

8   SHIPBOARD AND OTHER RANKS

1 Livy, VIII.8, gives the number as sixty men

2 Livy, VIII.8

3 From inventories from Piraeus naval base, inscribed on stone tablets and dating from the late fourth century BC. Listed in *Inscriptiones Graecae* 2/2. 1604–1632

4 Livy, XXI.49, mentions a foray by twenty Roman quinqueremes with 1,000 troops, i.e. fifty per ship, but this was for a raid on the enemy coast.

5 The foremast of an ancient ship carried an *artemo* or *artemon*, sail; the word is very rare, however. The earliest depiction of a two-masted ship is Etruscan.

6 Polybius, III.22/25/27; Livy, VII.27

7 There is a grave stele of such a rank from Ravenna, of the first century AD

8 Warry, *Warfare in the Classical World*; Polybius VI.24; Livy, VIII.8

9 The more permanent naval bases of this period have attendant cemeteries which have yielded many grave stelae of naval personnel

10 Starr, *The Roman Imperial Navy*; Sprague de Camp, *The Ancient Engineers*

9   RECRUITMENT

1 Scullard, *A History of the Roman World*

2 Livy I.43; Polybius VI.19 et seq.

3 The size and nature of military contribution was fixed by treaty between Rome and each city or other entity individually

4 Scullard, *A History of the Roman World*

5 Livy XXVI.47

6 Suetonius, *Augustus* 16

7 Salmon, *Samnium and the Samnites*

8 Polybius I.49

9 Lvy XXX.2

10 Livy XII.11

11 Ormerod, *Piracy in the Ancient World*

12 Scullard, *A History of the Roman World*.

13 Letters home from Egyptian recruits have survived

14 Starr, *The Roman Imperial Navy*

10   TERMS OF SERVICE

1 Roman money had four *quadrans* to one *as*, four (originally ten) *asses* to one *sestertius*, four *sestertii* to one *denarius* and twenty-five *denarii* to one gold *aureus*. In the east, Greek denominations were used; there were six *obols* to one *drachma*, one hundred *drachmae* to one *mina* and sixty *minae* to one *talent*. Cowell, *Everyday Life in Ancient Rome*.

2 Such relationships were in fact 'unofficial' until AD 19

3 Scullard, *From the Gracchi to Nero*

4 Starr, *The Roman Imperial Navy*

11   TRAINING

1 Polybius I.20

2 Based on the known crew for an Athenian trireme of 16 sailors and 170 rowers per ship and for the quinquereme, estimated crews of 20 sailors and 296 rowers per ship.

3 Morrison, Coates and Rankov, *The Athenian Trireme* and Welsh, *Building the Trireme*, for accounts of recent experience in the training and operation of a trireme crew.

4 Polybius I.21. The method of erecting rigs and training rowing crews ashore was older, being recorded in use for Egyptian recruits in the fourth century BC: Polyaenus III.11

5 Herodotus VI.12

6 Pitassi, *The Navies of Rome*

7 Polybius I.59

8 Livy XXIII.40

9 Polybius X.46

10 Tacitus, *The Annals* IV.4

12   UNIFORMS AND WEAPONS

1 Welsh, *Building the Trireme*

2 Vegetius IV.37

3 Livy II.49

4 *The Augustan History, Clodius Albinus* 2

5 Part of a frieze in the National Archeological Museum, Naples

6 Vegetius IV.44

7 Ibid.

8 Caesar, *Gallic Wars* V.48; also Bishop and Coulston, *Roman Military Equipment*

9 Caesar, *The Battle for Gaul* III.14

10 Vegetius, *Epitome of Military Science* IV.46

11 Bishop and Coulston, *Roman Military Equipment*; Feugere, *Weapons of the Romans*

13   FOOD AND DRINK

1 Livy XXIX.36

2 Pliny the Elder, *Natural History* II.18, as to salt fish, XV.4 as to preserving olives; Apicius, *De Re Coquinaria*, Book I includes many methods of preserving foodstuffs, for example, recipe XX relates to the keeping of figs, apples, plums pears and cherries; IX, as to pork skin and other meats; X, how to reconstitute salted meat.

3 Shephard S., *Pickled, Bottled and Canned, The Story of Food Preserving*

4 Morrison, Coates and Rankov, *The Athenian Trireme*; Welsh, *Building the Trireme*

5 Pryor, *The Geographical Conditions of Galley Navigation in the Mediterranean*, in *The Age of the Galley*

6 A later account of a ship in a storm complains that 'the tank in the hold which held the drinking water broke and spilled the water' suggesting that the container was terracotta or cement-lined wood. Casson, *Travel in the Ancient World*

7 Livy, XXIV.11

8 Livy XXI.49

9 Livy XXVIII.45

10 Tacitus, *Agricola* XXV

11 Livy XXIX.25
12 Livy XXV.31
13 Livy XXVII.5
14 Livy III.27 refers to an infantryman carrying five days' bread ration.
15 Casson L., *Travel in the Ancient World*
16 Apicius, *De Re Coquinaria;*
17 Polybius VI.39
18 Athenaeus III.110–114
19 Tannahill R., *Food in History*
20 Pliny the Elder, *Natural History* XVIII.14
21 Thucydides, III.49
22 Casson L., *Ships and Seamanship*
23 Thucydides III.49

14   ON BOARD SHIP

1 Morrison, Coates and Rankov, *The Athenian Trireme*
2 Or 'supplementary rowers'? Livy XXVI.47
3 Livy XXIX.25, reports an admonishment to the troops on a fleet for the invasion of Africa in 204 BC by Scipio, not to interfere with the sailors. Also Tacitus, *Annals* II.21, as to what happens if they tried to help in an emergency (in the example, disaster)
4 Caesar, *Civil War* III.1.5
5 Livy XXVII.32
6 Livy XXVIII.46
7 Livy XXIV.36
8 Morrison, Coates and Rankov, *The Athenian Trireme*
9 As examples, Polybius reports (XVI) the loss of a ship with all hands at the battle of Chios in 201 BC; in the civil wars, at the second battle between Caesar's commander Brutus and the ships of Marseilles (supporting Pompeius), one of the latter's smaller ships capsized when the troops on board rushed to one side and the rowers below were trapped and drowned (Caesar, *Civil Wars*, II.I.6 and 7)
10 Casson L., *Ships and Seamanship*
11 Pitassi, *The Navies of Rome*
12 Orosius VI.19
13 Livy XXIX.25

15   RELIGION AND SUPERSTITION

1 Names such as Diana, Juno, Jupiter, Ops, Isis and Minerva are attested for warships from grave stelae
2 Barker and Rasmussen, *The Etruscans*
3 Shepard, *Sea Power in Ancient History*
4 Tacitus, *The Histories* II.4
5 Guerber, *Myths of Greece and Rome*
6 *Larousse Encyclopaedia of Mythology*
7 *Larousse Encyclopaedia of Mythology*
8 Bickerman E.vJ., *Chronology of the Ancient World*
9 Marlow, *The Golden Age of Alexandria*
10 Tacitus, *Germania* IX
11 For example, Venice's annual 'Wedding with the Sea' and similar ceremonies where saints effigies are taken into or upon the sea in Italy, Greece and Spain at the opening of the traditional season
12 *Larousse Encyclopaedia of Mythology*

13 Now in the National Archaeological Museum, Naples

16   NAVIGATION

1 Pryor, *The Geographical Conditions of Galley Navigation in the Mediterranean*, in Morrison, *The Age of the Galley;* also Tacitus, *The Histories* II.98
2 Vegetius, *Epitome of Military Science* IV.38, codifying earlier writers
3 Vegetius, *Epitome of Military Science* IV.39; also Livy, V.2
4 Livy XXVI.41, 'In Spain at the beginning of spring, Scipio got his ships afloat again'; also Vegetius, *Epitome of Military Science* IV.39
5 Vegetius, *Epitome of Military Science* IV.39
6 Caesar, *Civil War* III.1.6 and III.3.25, where Caesar urges his captains to sail regardless of the winter
7 Acts 27.9; also Tacitus, *Histories* IV.38
8 Tacitus, *Agricola* XXIX
9 Pryor, *The Geographical Conditions of Galley Navigation in the Mediterranean*
10 Tacitus, *The Annals*, XV.46
11 Bradford, *Ulysses Found*; the author has stood on the coast of Cilento and seen a clear, calm day turn into a raging squall within twenty minutes
12 Duncan, *The Calendar*
13 James and Thorpe, *Ancient Inventions*
14 Cunliffe, *The Extraordinary Voyage of Pytheas the Greek*
15 James and Thorpe, *Ancient Inventions*
16 Duncan, *The Calendar*
17 Vegetius, *Epitome of Military Science* IV.42
18 Casson, *The Ancient Mariners*
19 Ibid.
20 Cicero, *Letters to Atticus*
21 James and Thorpe, *Ancient Inventions*
22 Sprague de Camp, *The Ancient Engineers*
23 Selkirk, *The Piercebridge Formula*
24 Ibid.
25 Sobel, *Longitude*
26 Marlowe, *The Golden Age of Alexandria*
27 It includes an inscription recording that an altar was set up by a prefect of the *Classis Britannica*; Rogan, *Reading Roman Inscriptions*
28 Sprague de Camp, *The Ancient Engineers*; James and Thorpe, *Ancient Inventions*
29 Baines and Malek, *Atlas of Ancient Egypt*
30 A. H. M. Jones, *Augustus*
31 Tacitus, *The Annals* XV.41
32 Tacitus, *The Annals* XIII.51
33 The canal was completed in 1893, using Nero's surveyed route; Sprague de Camp, *The Ancient Engineers*
34 Tacitus, *The Annals* XI.19
35 Selkirk, *The Piercebridge Formula*

17   SHIP PERFORMANCE

1 A knot is a rate of travel of one nautical mile (6,080 feet, 1,853 m) per hour, equivalent to 1.15 mph (1.85 kph)
2 Thucydides III.49

3 Rodgers, *Greek and Roman Naval Warfare*
4 Morrison, Coates and Rankov, *The Athenian Trireme*
5 Rankov, *Sailing into the Past*
6 Haywood, *Dark Age Naval Power*; Johansen, *The Viking Ships of Skuldelev*, in *Sailing into the Past*
7 Rodgers, *Greek and Roman Naval Warfare*
8 Pitassi, *Roman Warships*
9 It must be emphasised that there is no data as to the performance of late Roman seagoing ships and that these observations can only be based on consideration of the limited knowledge of the ships and of the relative technology
10 Thucydides, III.49, mentions of the crew of a trireme 'that they continued rowing while they ate their barley meal kneaded with wine and oil and they slept and rowed in turns'
11 Livy XXIX.27
12 Caesar, *Civil War* II.III.23
13 Caesar, *Bello Africanum* II
14 Livy XLV.41
15 Thucydides VI
16 Xenophon, *The Persian Expedition* VI.4; Morrison, Coates and Rankov, *The Athenian Trireme*
17 Thucydides, III.49
18 Casson, *Ships and Seamanship* contains a wide selection of sample voyages
19 Livy, XXII.11 and XXII.31
20 Livy, XXVII.5
21 Procopius, *The Vandal Wars* I.25.21
22 Morrison and Coates, *Greek and Roman Oared Warships*
23 Livy, XXVIII.30, where he says that the quinquereme was slower than a trireme
24 Morrison and Coates, *Greek and Roman Oared warships*

18 BASES AND SHORE ESTABLISHMENTS

1 Livy, XXVII.22
2 Livy, XXIX.25. Also Livy, III.26, referring to the existence of shipyards on the Tiber in his time
3 Tacitus, *The Annals* IV.4
4 The cistern remains largely intact and its sheer size indicates that a large number of ships could be 'watered'; also it provided a reserve for the base if the aqueduct was interrupted
5 Norwich, *Byzantium*
6 Norwich, *A History of Venice*
7 As the fleet did, for example, in AD 106, to assist in suppressing unrest in the Levant, Egypt and Judaea
8 Tacitus, *The Annals* I.8
9 Bounegru and Zahariade, *Les Forces Navales*
10 Tacitus, *The Annals* XII.14

19 STRATEGY AND DEPLOYMENT

1 Polybius III.22
2 Sitwell, *The World the Romans Knew*
3 Cunliffe, *The Extraordinary Voyage of Pytheas the Greek*
4 Harden, *The Phoenicians*
5 Casson, *Libraries in the Ancient World*
6 James and Thorpe, *Ancient Inventions*
7 As further examples, Tacitus, *Agricola* X, for the geography of Britain; *Agricola* XXIV, for that of Ireland; also see chapter 16
8 Tacitus, *Agricola* X and XXXVIII
9 Livy, XXIV.11
10 Pitassi, *The Navies of Rome*; Polybius, II.11
11 Livy, XXI.17
12 Livy, XXVII.22
13 Livy, XXIX.25
14 Pitassi, *The Navies of Rome*
15 For an examination of the progression of Roman military strategy in these areas, see Luttwak, *The Grand Strategy*
16 *Codex Theodosianus* VII.17

20 SCOUTING AND INTELLIGENCE

1 Austin and Rankov, *Exploratorio, Military and Political Intelligence in the Roman World etc.*
2 Polybius, I. 60 and 61
3 Livy, XXIII. 34 and 39
4 Plutarch, *Lives, Marcus Cato*, XXVII
5 Livy, XXIII. 34 and 38; Polybius, VII.9
6 Caesar, *The Battle for Gaul*, IV.21
7 Pitassi, *Roman Warships*
8 Austin and Rankov, *Exploratorio*
9 Morrison and Coates, *Greek and Roman Oared Warships*; Morrison, Coates and Rankov, *The Athenian Trireme*
10 Livy, XXII.19
11 The type clearly had an advantage, almost certainly in speed, as the Romans adopted it and it replaced all the earlier light warship types
12 Polybius, I.36
13 Pitassi, *Roman Warships*
14 Vegetius, *Epitome of Military Science* IV.37

21 BLOCKADES

1 The basic concept of Alfred Thayer Mahon's theories of naval strategy
2 Pompeius' admiral, Bibulus died 'worn out by hard work' in blockading the southern Adriatic in 48 BC, Dio, *Roman History* XLI.48; Caesar, *Civil War*, III.I.18
3 The great exception was the destruction of the Athenian fleet at Aegospotami in 405 BC, without which her enemies could stop the food carrying ships coming through the Dardanelles, upon which she depended
4 Guilmartin, *Gunpowder and Galleys*, for an examination of the problem of trying to apply Mahan's theories to galley warfare
5 Caesar, *The Battle for Gaul*, III.II.12
6 An example was Cicero's leisurely journey across the Aegean
7 Livy, XXIX.25
8 Shepard, *Sea Power in Ancient History*
9 Caesar, *Civil War*, III.III.23/24; Cassius Dio, *Roman History*, XII.48
10 Caesar, *Civil War*, III.I.9
11 Thucydides, *The Peloponnesian Wars*, VII.12, writing of the Athenian ships at the siege of Syracuse

12 Bradford, *Ulysses Found,* recounting personal experience of sailing in these waters

13 Polybius I.42

14 It was in fact to this ship, that the Romans copied, that Polybius refers.

15 The remains of these two ships have been recovered and are now at Marsala. They are identified as Punic from their markings but analysis of the stone aboard them shows it to have come from quarries on mainland Italy

16 Polybius, I.11

17 Caesar, *Civil War,* III.I.15 and 17

18 Cassius Dio, *Roman History,* XII.48; Caesar, *Civil War,* III.I 18

19 Caesar, *Civil War,* III.III.26, 27 and 28

20 Severin, *The Ulysses Voyage;* Pitassi, *The Navies of Rome*

22   BATTLE TACTICS

1 As illustrated by the relief of Rameses III victory over the Sea Peoples in 1176 BC, on his mortuary temple at Medinet Habu

2 The earliest-known clear representation of a ram, an engraving on a bronze fibula found at Athens and dated to about 850 BC.

3 For example, Polybius XVI.3, '. . being carried on past the enemy, lost his starboard oars and the timbers supporting his rowers were shattered . . .', detail from a description of a battle off Chios in 201 BC, between Macedonian and the Pergamene and Rhodian fleets.

4 Polybius XVI.3 as before and Thucydides II.84

5 Guilmartin, *Galleons and Galleys*

6 As an example of the confused fighting, Livy XXVI.39

7 Thucydides II.83

8 Polybius I.27and28; Pitassi, *The Navies of Rome*

9 Polybius as above

10 Bennett, *The Battle of Jutland*

11 Guilmartin, *Gunpowder and Galleys*

23   THE PERIPLOUS BATTLE

1 Although the later translation of Polybius refers to this ship as 'rowed seven men to an oar', i.e. a monoreme, this is probably by reference to Renaissance rowing systems; this ship is more likely to have been an enlarged sexteres with an additional thranite rower. The ship was in fact a Hellenistic type, having been captured by the Carthaginians from Pyrrhus of Epirus in 276 BC (Polybius I.23).

2 Silius Italicus, XIV, writing in the first century BC; Pliny, *Natural History,* XXXII.1

3 Polybius II.23.

4 Appian, *The Civil Wars.*

24   THE DIEKPLOUS BATTLE

1 Livy, Book XXXVII.29/30 and Appian, *Syrian Wars,* are the ancient sources; also Rodgers, *Greek and Roman Naval Warfare.*

2 This was an iron basket, filled with blazing material which seems to have been unique to the Rhodians. This device had been used by them, slung either side of the bows, to escape from a trap at the harbour of Panormus earlier in the year. There was a natural antipathy to fire on these wooden ships and it was not generally used as a weapon, being as dangerous for the one using it as for their intended victim.

3 The Roman writer Florus describing Antonius' ships, quoted in Shepard, *Sea Power in Ancient History*

4 Appian, *The Civil Wars* IV and V; Pitassi, *The Navies of Rome*

5 Orosius VI.19, the mechanics of this have been considered above

6 Cassius Dio L.32; this is the only time that the use of such weapons in a sea battle is attested, although common on land, they were generally avoided onboard ship for obvious reasons.

25   THE INFLUENCE OF TOPOGRAPHY

1 Polybius I. 24

2 Polybius I.27/ 28

3 Polybius I.49/ 50/ 51

4 Polybius I.50

26   THE STATIC BATTLE

1 Polybius XIV.2 and Livy XXX.10

2 Appian, *The Civil Wars* V

27   RHODES

1 Burn, *The Pelican History of Greece*

2 Herodotus VII.94/95; *Nelson, The Battle of Salamis*

3 Herodotus VIII.130

4 Herodotus IX.90–105

5 Thucydides VI.43

6 Berthold, *Rhodes in the Hellenistic Age*

7 Diodorus Siculus XVI.7

8 Berthold, *Rhodes in the Hellenistic Age*

9 Morrison and Coates, *Greek and Roman Oared Warships*

10 Polybius XXX.5

11 Diodorus Siculus, Book XX deals with the siege; Rodgers, *Greek and Roman Naval Warfare*

12 Polybius IV.47

13 Green, *Alexander to Actium;* Polybius V.89

14 Livy XXVII.30

15 Berthold, Rhodes in the Hellenistic Age

16 Polybius XVI.2–16; probably the most detailed near-contemporary description of an ancient naval battle to have survived

17 Livy XXXVII.9

18 Livy XXXVII.22

19 Livy XXXVII.29/30; Polybius XXI.7

20 Caesar, *Civil War* III.5

21 Caesar, *Civil War* III.3.27

22 Caesar, *Civil War* III.5.106

23 Marlowe, *The Golden Age of Alexandria;* Hirtius (one of Caesar's legionary commanders) *The Alexandrine War*

24 Appian, *Civil War* IV.66

25 Berthold, *Rhodes in the Hellenistic Age*

26 Starr, *The Roman Imperial Navy*

28   PERGAMUM AND SYRACUSE

1 Green, *Alexander to Actium*
2 Rodgers, *Greek and Roman Naval Warfare*; Polybius XVI.3
3 Bickerman, *Chronology of the Ancient World*
4 Livy XXVIII.29
5 Livy XXVIII.8
6 Rodgers, *Greek and Roman Naval Warfare*
7 Polybius XVI contains a graphic description of this battle
8 Green, *Alexander to Actium*
9 Herodotus VII.155
10 Herodotus VII.158
11 Thucydides Book VII
12 Diodorus Siculus XIV.42
13 Diodorus Siculus XIV.51; Marsden, *Greek and Roman Artillery*
14 Casson, *Ships and Seamanship*
15 Athenaeus V
16 Livy XXI.49
17 Livy XXII.37
18 Champion, *The Tyrants of Syracuse*
19 Shepard, *Sea Power in Ancient History*
20 Livy XXIV.36
21 Livy XXIV.33/34; Polybius VIII.4/5

29   MARSEILLES AND JUDAEA

1 Herodotus, *The Histories* I.163
2 Herodotus, *The Histories* I.166
3 Champion, *The Tyrants of Syracuse*
4 Livy XXII.19
5 Caesar, *Civil War* I.III.56
6 Caesar, *Civil War* I.III.57 and 58
7 Caesar, *The Civil War* II.I.4
8 Caesar, *Civil War* II.I.6
9 Scullard, *A History of the Roman World*
10 Rogerson, *Chronicle of the Old Testament Kings*
11 Perowne, *The Life and Times of Herod the Great*
12 Patai, *The Children of Noah*

30   CARTHAGE, TARANTO AND THE PIRATES

1 Polybius III.22
2 Harden, *The Phoenicians*; Moscati, *The World of the Phoenicians*
3 Polybius III.24
4 Polybius III.25
5 Appian, VIII, Libyca 96
6 Harden, *The Phoenicians*
7 Polybius VI.52
8 Pitassi, *The Navies of Rome*
9 Ormerod, *Piracy in the Ancient World*
10 Berthold, *Rhodes in the Hellenistic Age*
11 Livy, VIII.3
12 Scullard, *A History of the Roman World*
13 Livy, XXV.10
14 Livy, XXVI.39

31   HELLENISTIC NAVIES

1 Casson, *Ships and Seamanship*
2 Caesar, *Civil War* III.1.5
3 Caesar, *Civil War* III.3.40
4 Caesar, *Civil War* III.5.111
5 Bickerman, *Chronology of the Ancient World*, for regnal dates
6 Polybius, XVII.3
7 Polybius, XVIII.44
8 Appian, *The Mithridatic Wars*
9 Pitassi, *The Navies of Rome*
10 Green, *Alexander to Actium*

32   THE BARBARIANS

1 Caesar, *The Battle for Gaul* III.13
2 Caesar, *The Battle for Gaul* III.14 and 15
3 Tacitus, *Germania* 44
4 Christensen, *Proto-Viking and Norse Craft* in Gardiner and Christensen, *The Earliest Ships*
5 Haywood, *Dark Age Naval Power*
6 Tacitus, *The Histories* V.23
7 The 'Hjortspring boat', Christensen, *Scandinavian Ships from earliest times to the Vikings*, in Bass, *A History of Seafaring*
8 Heather, *Empires and Barbarians*
9 Pearson, *The Roman Shore Forts*
10 Haywood, *Dark Age Naval Power*; Pearson, *The Roman Shore Forts*
11 Aurelius Victor, *De Caesaribus* 39
12 Ammianus Marcellinus, *Roman History* XXVI.5
13 Ammianus Marcellinus, *Roman History* XXVII.8
14 Tacitus, *The Histories* III.47
15 Ormerod, *Piracy in the Ancient World*
16 Starr, *The Influence of Seapower on Ancient History*
17 Pitassi, *The Navies of Rome*
18 Luttwak, *The Grand Strategy*
19 Aurelius Victor, *De Caesaribus* 33
20 Haywood, *Dark Age Naval Power*, quoting Orosius VII.41
21 Zosimus, I.42
22 Procopius, *The Vandalic Wars* III.3
23 Procopius, *The Vandalic Wars* III.4
24 Newark, *The Barbarians*
25 Procopius, *The Vandalic Wars* III.5
26 Procopius, *The Vandalic Wars* III.6

INDEX

181